Excel 2010 for Educational and Psychological Statistics

Thomas Quirk

Excel 2010 for Educational and Psychological Statistics

A Guide to Solving Practical Problems

 Springer

Thomas Quirk
School of Business and Technology
Webster University
St. Louis, MO 63119, USA
quirkto@webster.edu

ISBN 978-1-4614-2070-5   e-ISBN 978-1-4614-2071-2
DOI 10.1007/978-1-4614-2071-2
Springer New York Dordrecht Heidelberg London

Library of Congress Control Number: 2011941800

Printed on acid-free paper

Springer is part of Springer Science+Business Media (www.springer.com)

This book is dedicated to the more than 3,000 students I have taught at Webster University's campuses in St. Louis, London, and Vienna; the students at Principia College in Elsah, Illinois; and the students at the Cooperative State University of Baden-Wuerttemburg in Heidenheim, Germany. These students taught me a great deal about the art of teaching. I salute them all, and I thank them for helping me to become a better teacher.

Preface

Excel 2010 for Educational and Psychological Statistics: A Guide to Solving Practical Problems helps anyone who wants to learn the basics of applying Excel's powerful statistical tools to their work situation or to their classes. If understanding statistics is not your strongest suit, you are not mathematically inclined, or you are wary of computers, then this is the book for you.

You will learn how to perform key statistical tests in Excel without being overwhelmed by statistical theory. This book clearly and logically shows how to run statistical tests to solve practical problems in education and psychology.

Excel is a widely available computer program for students, instructors, and managers in education and in business. It is also an effective teaching and learning tool for quantitative analyses in statistics courses. Its powerful computational ability and graphical functions make learning statistics much easier than in years past. However, this is the first book to showcase Excel's usefulness in teaching educational and psychological statistics. And it focuses exclusively on this topic in order to render the subject matter applicable and practical – and, easy to comprehend and apply.

Unique features of this book:

- Includes 163 color screen shots so you can be sure you are performing Excel steps correctly.
- You will be told each step of the way, not only *how* to use Excel, but also *why* you are doing each step.
- Includes specific objectives embedded in the text for each concept, so you can know the purpose of the Excel steps.
- You will learn both how to write statistical formulas using Excel and how to use Excel's drop-down menus that will create the formulas for you.
- Statistical theory and formulas are explained in clear language without bogging you down in mathematical fine points.
- Practical examples of problems are taken from both education and psychology.

- Each chapter presents key steps to solve practical problems using Excel. In addition, three practice problems at the end of each chapter enable you to test your new knowledge. Answers to these problems appear in Appendix A.
- A "Practice Test" is given in Appendix B to test your knowledge at the end of the book. Answers to this test appear in Appendix C.
- This book does not come with a CD of Excel files which you can upload to your computer. Instead, you will be shown how to create each Excel file yourself. In a work or classroom situation, your colleagues and professors will not give you an Excel file. You will be expected to create your own. This book will give you ample practice in developing this important skill.
- This book is a tool that can be used either by itself or along with *any* good statistics book.

This book is appropriate for use in any course – graduate of undergraduate – in Educational and Psychological Statistics, as well as for administrators/managers who want to improve their Excel skills. It will also benefit students who are taking courses in Sociology, Anthropology, or Computer Science who want to learn how to use Excel to solve statistics problems.

The ideas in this book have been thoroughly tested by its author, Professor Tom Quirk, in both Marketing Statistics and Marketing Research courses.

At the beginning of his academic career, Prof. Quirk spent 6 years in educational research at The American Institutes for Research and Educational Testing Service. He then taught Social Psychology, Educational Psychology, and General Psychology at Principia College and is currently a Professor of Marketing in the George Herbert Walker School of Business & Technology at Webster University based in St. Louis, Missouri (USA) where he teaches Marketing Statistics, Marketing Research, and Pricing Strategies. He has published articles in the *Journal of Educational Psychology, Journal of Educational Research, Review of Educational Research, Journal of Educational Measurement, Educational Technology, The Elementary School Journal, Journal of Secondary Education, Educational Horizons, and Phi Delta Kappan*. In addition, he has written 60+ textbook supplements in Marketing and Management, published 20+ articles in professional journals, and presented 20+ papers at professional meetings, including annual meetings of The American Educational Research Association, The American Psychological Association, and the National Council on Measurement in Education. He holds a BS in Mathematics from John Carroll University, both an MA in Education and a PhD in Educational Psychology from Stanford University, and an MBA from The University of Missouri-St. Louis.

St. Louis, MO, USA   Thomas Quirk

Acknowledgments

Excel 2010 for Educational and Psychological Statistics: A Guide to Solving Practical Problems is the result of inspiration from three important people: my two daughters and my wife. Jennifer Quirk McLaughlin invited me to visit her MBA classes several times at the University of Witwatersrand in Johannesburg, South Africa. These visits to a first-rate MBA program convinced me there was a need for a book to teach students how to solve practical business problems using Excel. Meghan Quirk-Horton's dogged dedication to learning the many statistical techniques needed to complete her PhD dissertation illustrated the need for a statistics book that would make this daunting task more user-friendly. And Lynne Buckley-Quirk was the number one cheerleader for this project from the beginning, always encouraging me and helping me remain dedicated to completing it.

Sue Gold, a reference librarian at Webster University in St. Louis, was a valuable colleague in helping me to do key research, and was a steady supporter of this idea. Brad Wolaver of Webster University improved my Office 2010 skills in many ways.

Marc Strauss, my editor at Springer, caught the spirit of this idea in our first phone conversation and shepherded this book through the idea stages until it reached its final form. His encouragement and support were vital to this book seeing the light of day. I thank him for being such an outstanding product champion throughout this process.

Contents

Chapter 1
Sample Size, Mean, Standard Deviation, and Standard Error of the Mean

This chapter deals with how you can use Excel to find the average (i.e., "mean") of a set of scores, the standard deviation of these scores (STDEV), and the standard error of the mean (s.e.) of these scores. All three of these statistics are used frequently and form the basis for additional statistical tests.

1.1 Mean

The *mean* is the "arithmetic average" of a set of scores. When my daughter was in the fifth grade, she came home from school with a sad face and said that she did not get "averages." The book she was using described how to find the mean of a set of scores, and so I said to her:

"Jennifer, you add up all the scores and divide by the number of numbers that you have."

She gave me "that look," and said: "Dad, this is serious!" She thought I was teasing her. So, I said:

"See these numbers in your book; add them up. What is the answer?" (She did that).

"Now, how many numbers do you have?" (She answered that question).

"Then, take the number you got when you added up the numbers, and divide that number by the number of numbers that you have."

She did that and found the correct answer. You will use that same reasoning now, but it will be much easier for you because Excel will do all of the steps for you.

We will call this average of the scores the "mean" which we will symbolize as $\bar{X}$, and we will pronounce it as "Xbar."

The formula for finding the mean with your calculator looks like this:

$$\bar{X} = \frac{\sum X}{n} \tag{1.1}$$

T. Quirk, *Excel 2010 for Educational and Psychological Statistics:*
A Guide to Solving Practical Problems, DOI 10.1007/978-1-4614-2071-2_1,
© Springer Science+Business Media, LLC 2012

The symbol Σ is the Greek letter sigma, which stands for "sum." It tells you to add up all the scores that are indicated by the letter X and then to divide your answer by n (the number of numbers that you have).

Let us give a simple example.

Suppose that you had these six test scores on a seven-item true-false quiz:

6
4
5
3
2
5

To find the mean of these scores, you add them up and then divide by the number of scores. So, the mean is: $25/6 = 4.17$.

1.2   Standard Deviation

The *standard deviation* tells you "how close the scores are to the mean." If the standard deviation is a small number, this tells you that the scores are "bunched together" close to the mean. If the standard deviation is a large number, this tells you that the scores are "spread out" a greater distance from the mean. The formula for the standard deviation (which we will call STDEV and use the letter, S, to symbolize) is:

$$\text{STDEV} = S = \sqrt{\frac{\sum (X - \bar{X})^2}{n - 1}} \qquad (1.2)$$

The formula looks complicated, but what it asks you to do is this:

1. Subtract the mean from each score $(X - \bar{X})$.
2. Then, square the resulting number to make it a positive number.
3. Then, add up these squared numbers to get a total score.
4. Then, take this total score and divide it by $n - 1$ (where n stands for the number of numbers that you have).
5. The final step is to take the square root of the number you found in step 4.

You will not be asked to compute the standard deviation using your calculator in this book, but you could see examples of how it is computed in any basic statistics book. Instead, we will use Excel to find the standard deviation of a set of scores. When we use Excel on the six numbers we gave in the description of the mean above, you will find that the *STDEV* of these numbers, S, is 1.47.

1.3   Standard Error of the Mean

The formula for the *standard error of the mean, s.e.,* (which we will use $S_{\bar{X}}$ to symbolize) is

$$\text{s.e.} = S_{\bar{X}} = \frac{S}{\sqrt{n}} \tag{1.3}$$

To find *s.e.*, all you need to do is to take the standard deviation, STDEV, and divide it by the square root of *n*, where *n* stands for the "number of numbers" that you have in your data set. In the example under the standard deviation description above, the s.e. = 0.60. (You can check this on your calculator.)

If you want to learn more about the standard deviation and the standard error of the mean, see Weiers (2011).

Now, let us learn how to use Excel to find the sample size, the mean, the standard deviation, and the standard error or the mean using a geometry test given to a class of eight ninth graders at the end of the first term of the school year (50 points possible). The hypothetical data appear in Fig. 1.1.

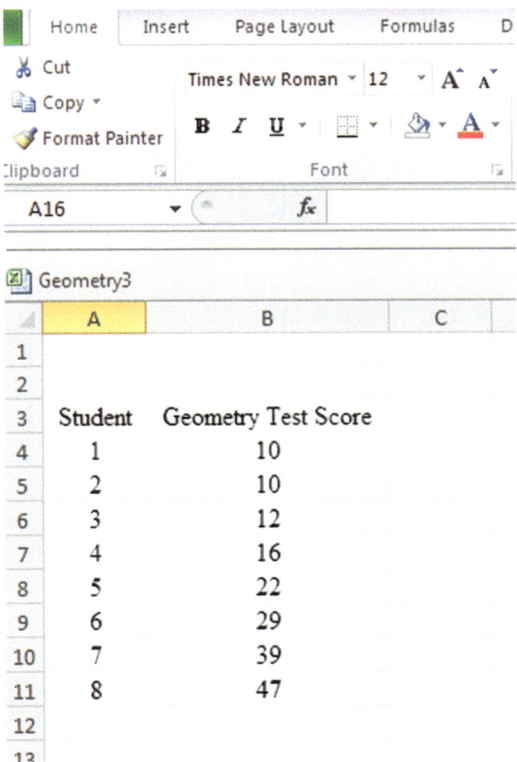

Fig. 1.1 Worksheet data for a geometry test (practical example)

1.4   Sample Size, Mean, Standard Deviation, and Standard Error of the Mean

> Objective: To find the sample size (n), mean, standard deviation (STDEV), and standard error of the mean (s.e.) for these data

Start your computer and click on the Excel 2010 icon to open a blank Excel spreadsheet.

Enter the data in this way:

A3:   Student
B3:   Geometry Test Score
A4:   1

1.4.1   Using the Fill/Series/Columns Commands

> Objective: To add the student numbers 2–8 in a column underneath student #1

Put pointer in A4.
Home (top left of screen)
Fill (top right of screen: click on the down arrow; see Fig. 1.2)

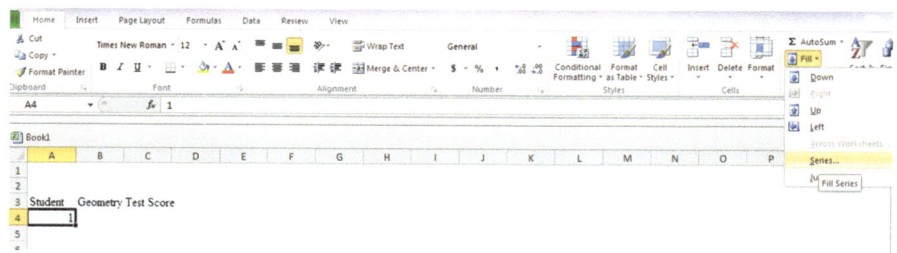

Fig. 1.2   Home/Fill/Series commands

Series
Columns
Step value: 1
Stop value: 8 (see Fig. 1.3)

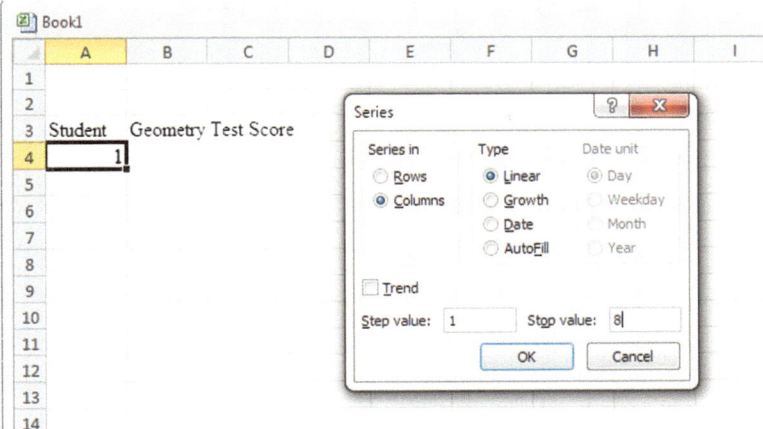

Fig. 1.3 Example of dialog box for Fill/Series/Columns/Step value/Stop value commands

OK.

The student numbers should be identified as 1–8, with 8 in cell A11.

Now, enter the Geometry Test Scores in cells B4:B11.

Since your computer screen shows the information in a format that does not look professional, you need to learn how to "widen the column width" and how to "center the information" in a group of cells. Here is how you can do those two steps.

1.4.2 Changing the Width of a Column

Objective: To make a column width wider so that all of the information fits inside that column

If you look at your computer screen, you can see that column B is not wide enough so that all of the information fits inside this column. To make column B wider

Click on the letter, B, at the top of your computer screen.

Place your mouse pointer at the far right corner of B until you create a "cross sign" on that corner.

Left click on your mouse, hold it down, and move this corner to the right until it is "wide enough to fit all of the data."

Take your finger off the mouse to set the new column width (see Fig. 1.4).

Fig. 1.4 Example of how to widen the column width

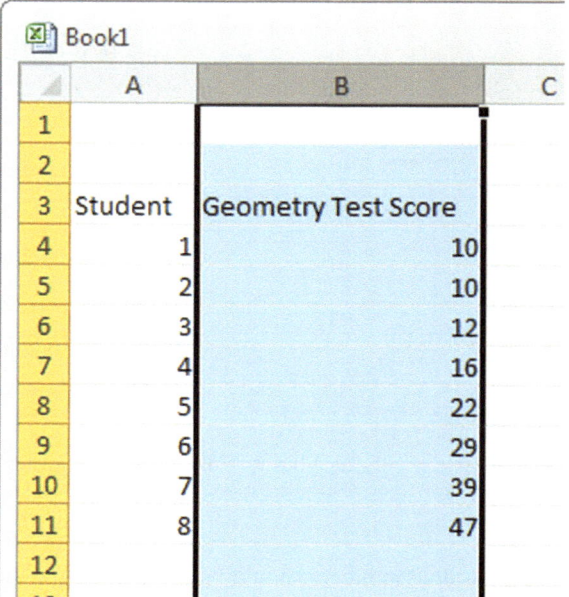

Then, click on any empty cell (i.e., any blank cell) to "deselect" column B so that it is no longer a darker color on your screen.

When you widen a column, you will make all of the cells in all of the rows of this column that same width.

Now, let us go through the steps to center the information in both column A and column B.

1.4.3 Centering Information in a Range of Cells

Objective: To center the information in a group of cells

In order to make the information in the cells look "more professional," you can center the information using the following steps:

Left click your mouse on A3 and drag it to the right and down to highlight cells A3: B11 so that these cells appear in a darker color.

At the top of your computer screen, you will see a set of "lines" in which all of the lines are "centered" to the same width under "alignment" (it is the second icon at the bottom left of the alignment box, see Fig. 1.5).

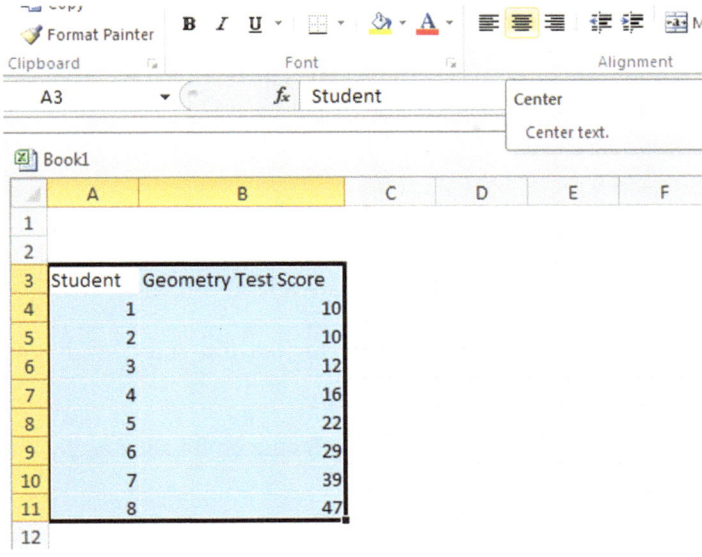

Fig. 1.5 Example of how to center information within cells

Click on this icon to center the information in the selected cells (see Fig. 1.6).

Fig. 1.6 Final result of centering information in the cells

Since you will need to refer to the geometry test scores in your formulas, it will be much easier to do this if you "name the range of data" with a name instead of having to remember the exact cells (B4:B11) in which these figures are located. Let us call that group of cells Geometry, but we could give them any name that you want to use.

1.4.4   Naming a Range of Cells

Objective: To name the range of data for the test scores with the name: Geometry

Highlight cells B4:B11 by left clicking your mouse on B4 and dragging it down to B11.
Formulas (top left of your screen)
Define Name (top center of your screen)
Geometry (type this name in the top box; see Fig. 1.7)

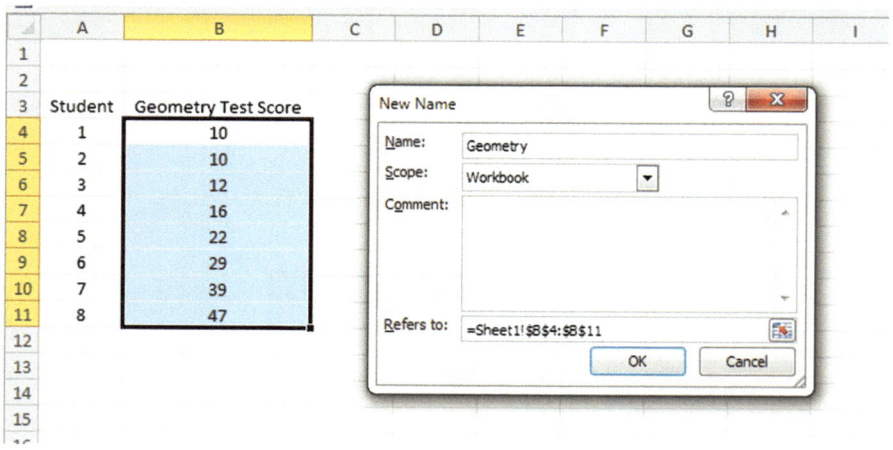

Fig. 1.7 Dialog box for "naming a range of cells" with the name: Geometry

OK.
Then, click on any cell of your spreadsheet that does not have any information in it (i.e., it is an "empty cell") to deselect cells B4:B11.
Now, add the following terms to your spreadsheet:

E6:   *n*
E9:   Mean
E12:   STDEV
E15:   s.e. (see Fig. 1.8)

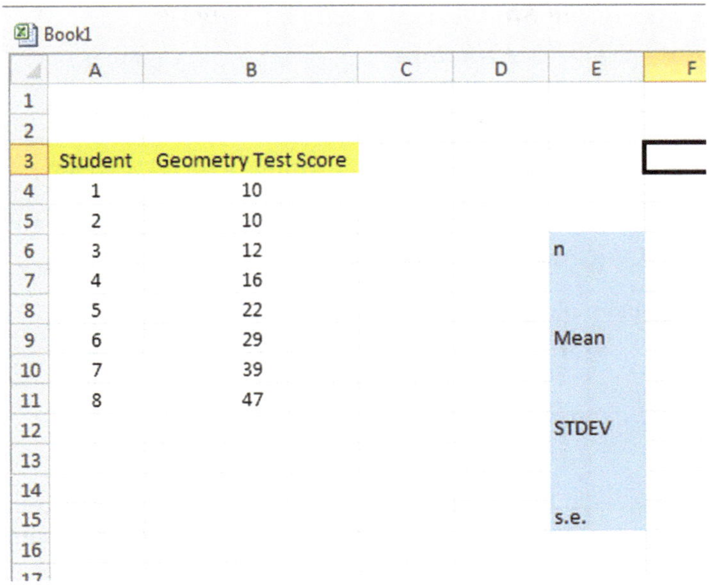

	A	B	C	D	E	F
1						
2						
3	Student	Geometry Test Score				
4	1	10				
5	2	10				
6	3	12			n	
7	4	16				
8	5	22				
9	6	29			Mean	
10	7	39				
11	8	47				
12					STDEV	
13						
14						
15					s.e.	
16						

Book1

Fig. 1.8 Example of entering the sample size, mean, STDEV, and s.e. labels

Note: Whenever you use a formula, you must add an equal sign (=) at the beginning of the name of the function so that Excel knows that you intend to use a formula.

1.4.5   Finding the Sample Size Using the =COUNT Function

Objective: To find the sample size (n) for these data using the =COUNT function

F6:   = COUNT (Geometry)

This command should insert the number 8 into cell F6 since there are eight students in this class.

1.4.6   Finding the Mean Score Using the =AVERAGE Function

Objective: To find the mean sales figure using the =AVERAGE function

F9:   = AVERAGE (Geometry)

This command should insert the number 23.125 into cell F9.

1.4.7   Finding the Standard Deviation Using the =STDEV Function

> Objective: To find the standard deviation (STDEV) using the =STDEV function

F12:   =STDEV (Geometry)

This command should insert the number 14.02485 into cell F12.

1.4.8   Finding the Standard Error of the Mean

> Objective: To find the standard error of the mean using a formula for these eight data points

F15:   =F12/SQRT (8)

This command should insert the number 4.958533 into cell F15 (see Fig. 1.9).

▣ Book1							
◢	A	B	C	D	E	F	G
1							
2							
3	Student	Geometry Test Score					
4	1	10					
5	2	10					
6	3	12			n	8	
7	4	16					
8	5	22					
9	6	29			Mean	23.125	
10	7	39					
11	8	47					
12					STDEV	14.02485	
13							
14							
15					s.e.	4.958533	
16							
17							

Fig. 1.9 Example of using Excel formulas for sample size, mean, STDEV, and s.e.

Important note: *Throughout this book, be sure to double-check all of the figures in your spreadsheet to make sure that they are in the correct cells, or the formulas will not work correctly!*

1.4.8.1   Formatting Numbers in Number Format (Two Decimal Places)

Objective: To convert the mean, STDEV, and s.e. to two decimal places

Highlight cells F9:F15.

Home (top left of screen)

Look under "number" at the top center of your screen. In the bottom right corner, gently place your mouse pointer on your screen at the bottom of the .00.0 until it says: "Decrease Decimals" (see Fig. 1.10).

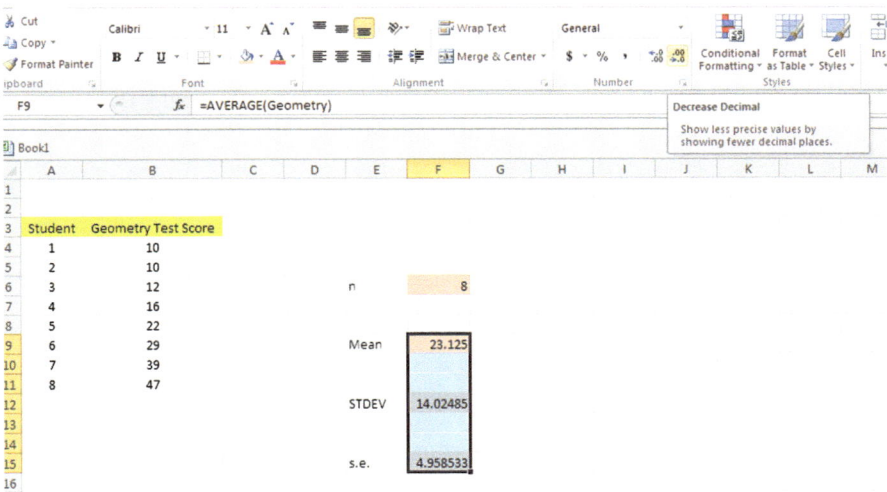

Fig. 1.10 Using the "Decrease Decimal" icon to convert numbers to fewer decimal places

Click on this icon *once* and notice that the cells F9:F15 are now all in just two decimal places (see Fig. 1.11).

◢	A	B	C	D	E	F	G
1							
2							
3	Student	Geometry Test Score					
4	1	10					
5	2	10					
6	3	12			n	8	
7	4	16					
8	5	22					
9	6	29			Mean	23.13	
10	7	39					
11	8	47					
12					STDEV	14.02	
13							
14							
15					s.e.	4.96	
16							
17							

Fig. 1.11 Example of converting numbers to two decimal places

Now, click on any "empty cell" on your spreadsheet to deselect cells F9:F15.

1.5   Saving a Spreadsheet

Objective: To save this spreadsheet with the name: Geometry3

In order to save your spreadsheet so that you can retrieve it sometime in the future, your first decision is to decide "where" you want to save it. That is your decision and you have several choices. If it is your own computer, you can save it onto your hard drive (you need to ask someone how to do that on your computer). Or, you can save it onto a "CD" or onto a "flash drive." You then need to complete these steps:

File
Save as

> (*select the place where you want to save the file by scrolling either down or up the bar on the left and click on the place where you want to save the file, for example, Documents: My Documents location*)

File name: Geometry3 (enter this name to the right of file name; see Fig. 1.12)

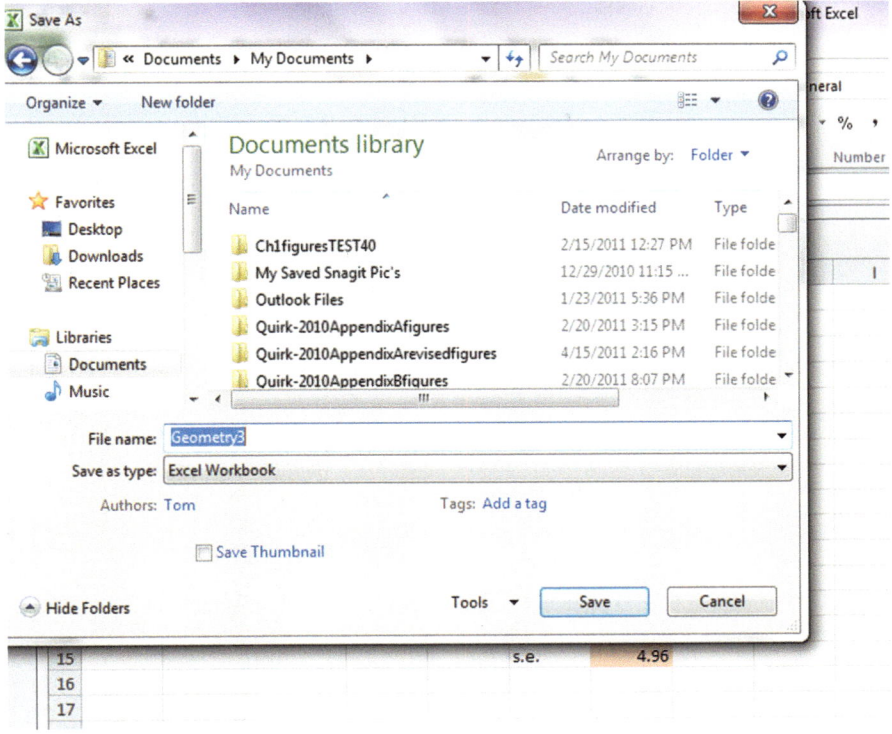

Fig. 1.12 Dialog box of saving an Excel workbook file as "Geometry3" in Documents: My Documents location

Save

Important note: *Be very careful to save your Excel file spreadsheet every few minutes so that you do not lose your information!*

1.6   Printing a Spreadsheet

Objective: To print the spreadsheet

Use the following procedures when printing any spreadsheet:

File
Print
Print Active Sheets (see Fig. 1.13)

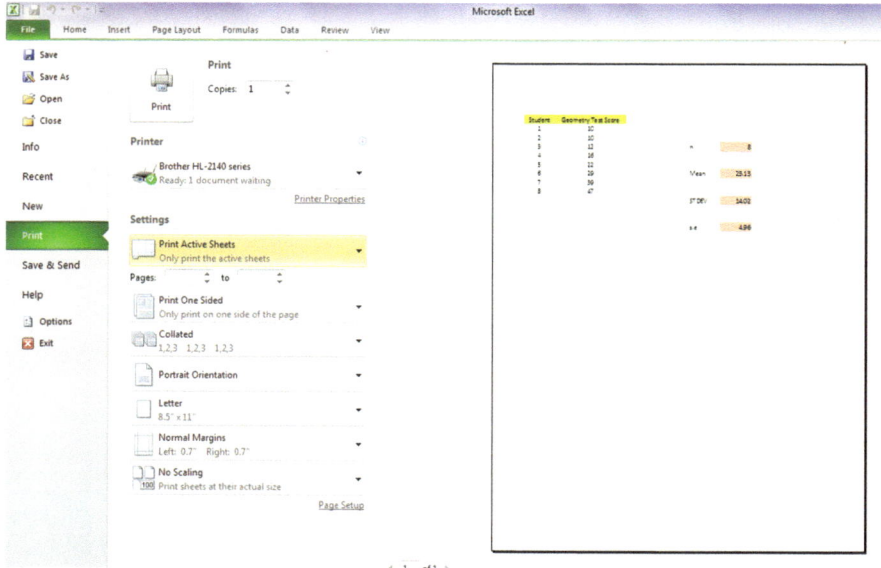

Fig. 1.13 Example of how to print an Excel worksheet using the File/Print/Print Active Sheets commands

Print (top of your screen)

The final spreadsheet is given in Fig. 1.14.

	A	B	C	D	E	F	G
1							
2							
3	Student	Geometry Test Score					
4	1	10					
5	2	10					
6	3	12			n	8	
7	4	16					
8	5	22					
9	6	29			Mean	23.13	
10	7	39					
11	8	47					
12					STDEV	14.02	
13							
14							
15					s.e.	4.96	
16							
17							

Fig. 1.14 Final result of printing an Excel spreadsheet

Before you leave this chapter, let us practice changing the format of the figures on a spreadsheet with two examples: (1) using two decimal places for figures that are dollar amounts and (2) using three decimal places for figures.

Close your spreadsheet by using File/Close and open a blank Excel spreadsheet by using File/New/Create (on the far right of your screen).

1.7   Formatting Numbers in Currency Format (Two Decimal Places)

> Objective: To change the format of figures to dollar format with two decimal places

A3:   Price
A4:   1.25
A5:   3.45
A6:   12.95

Home.
Highlight cells A4:A6 by left clicking your mouse on A4 and dragging it down so that these three cells are highlighted in a darker color.
Number (top center of screen: click on the down arrow on the right; see Fig. 1.15)

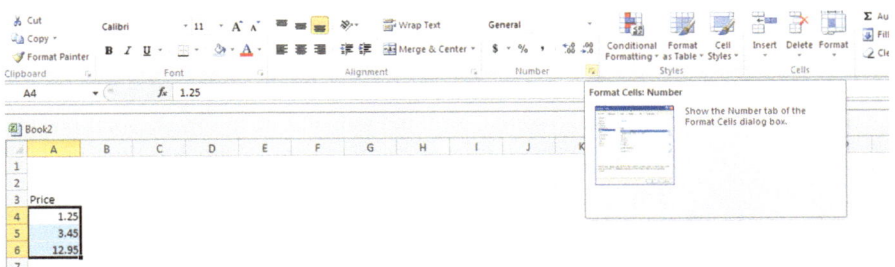

Fig. 1.15   Dialog box for number format choices

Category: Currency
Decimal places: 2 (then see Fig. 1.16)

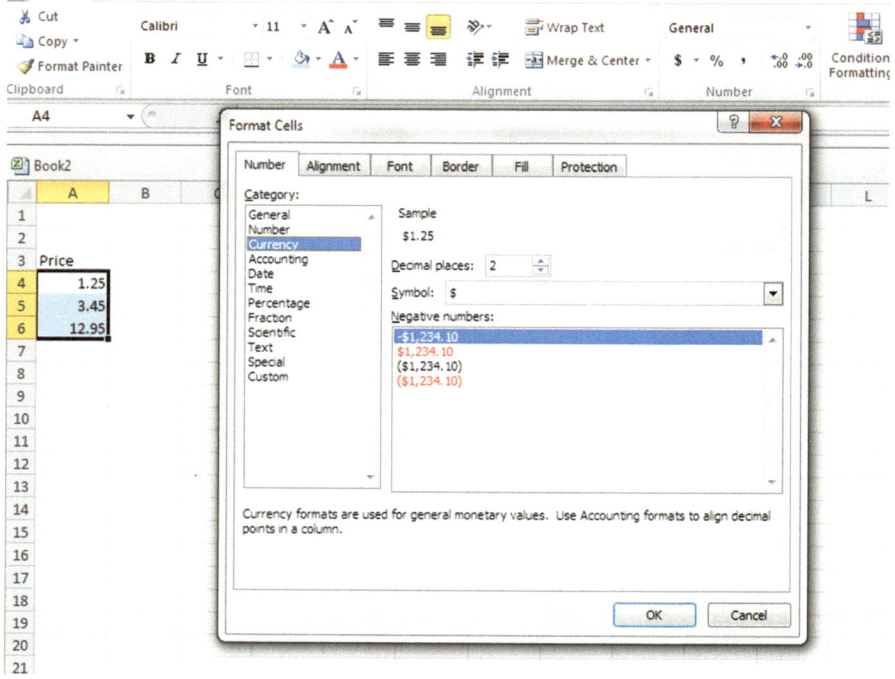

Fig. 1.16 Dialog box for currency (two decimal places) format for numbers

OK.

The three cells should have a dollar sign in them and be in two decimal places. Next, let us practice formatting figures in number format, three decimal places.

1.8   Formatting Numbers in Number Format (Three Decimal Places)

Objective: To format figures in number format, three decimal places

Home.
Highlight cells A4:A6 on your computer screen.
Number (click on the down arrow on the right)
Category: Number
At the right of the box, change two decimal places to three decimal places by clicking on the "up arrow" once.
OK.

The three figures should now be in number format, each with three decimals.

Now, click on any blank cell to deselect cells A4:A6. Then, close this file by File/
Close/Don't Save (since there is no need to save this practice problem).

You can use these same commands to format a range of cells in percentage
format (and many other formats) to whatever number of decimal places you want to
specify.

1.9 End-of-Chapter Practice Problems

1. Suppose that a fourth grade language arts teacher at Zach White School in El
   Paso, Texas, administers a 25-item test at the end of Chaps. 1–5 of the book:
   Charlotte's Web to test her students' reading comprehension of these chapters.
   The hypothetical data appear in Fig. 1.17:

Zach White School
4th grade unit on *Charlotte's Web*

Chapters 1-5 test (25 items)
20
18
19
16
23
24
10
12
11
8
16
14
15

Fig. 1.17 Worksheet data for
Chap. 1: practice problem #1

(a) Use Excel to the right of the table to find the sample size, mean, standard
   deviation, and standard error of the mean for these data. Label your answers
   and round off the mean, standard deviation, and standard error of the mean to
   two decimal places; use number format for these three figures.
(b) Print the result on a separate page.
(c) Save the file as: CHARLOTTE2.

2. Suppose that the Human Resources department of your company has administered a "morale survey" to all middle-level managers and that you have been asked to summarize the results of the survey. You have decided to test your Excel skills on one item to see if you can do this assignment correctly, and you have selected item #21 to test out your skills. The data are given in Fig. 1.18:

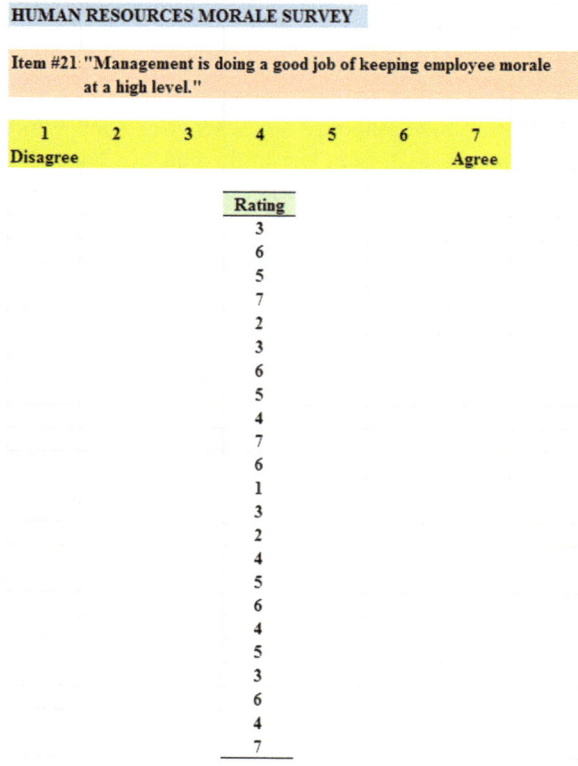

Fig. 1.18 Worksheet data for Chap. 1: practice problem #2

(a) Use Excel to create a table of these ratings, and at the right of the table use Excel to find the sample size, mean, standard deviation, and standard error of the mean for these data. Label your answers and round off the mean, standard deviation, and standard error of the mean to two decimal places using number format.

(b) Print the result on a separate page.

(c) Save the file as: MORALE4.

3. Suppose that a fifth grade science teacher at Deer Creek Elementary School in Bailey, Colorado, is using a textbook based on basic geology that typically requires about eight class days to teach each chapter. At the end of Chap. 8,

the teacher gives a 15-item true-false quiz on this chapter. The test results are given in Fig. 1.19:

Deer Creek Elementary School

5th grade science test

Chapter 8 (15 items)
12
15
13
8
10
12
13
12
9
4
11
15
13
15
12
14

Fig. 1.19 Worksheet data for Chap. 1: practice problem #3

(a) Use Excel to create a table for these data, and at the right of the table, use Excel to find the sample size, mean, standard deviation, and standard error of the mean for these data. Label your answers and round off the mean, standard deviation, and standard error of the mean to three decimal places using number format.
(b) Print the result on a separate page.
(c) Save the file as: SCIENCE8.

Reference

Weiers, R.M. Introduction to Business Statistics (7th ed.). Mason, OH: South-Western Cengage Learning, 2011.

Chapter 2
Random Number Generator

Suppose that a local school superintendent asked you to take a random sample of 5 of an elementary school's 32 teachers using Excel so that you could interview these five teachers about their job satisfaction at their school.

To do that, you need to define a "sampling frame." A sampling frame is a list of people from which you want to select a random sample. This frame starts with the identification code (ID) of the number 1 that is assigned to the name of the first teacher in your list of 32 teachers in this school. The second teacher has a code number of 2, the third a code number of 3, and so forth until the last teacher has a code number of 32.

Since this school has 32 teachers, your sampling frame would go from 1 to 32 with each teacher having a unique ID number.

We will first create the frame numbers as follows in a new Excel worksheet.

2.1 Creating Frame Numbers for Generating Random Numbers

Objective: To create the frame numbers for generating random numbers

A3:   FRAME NO.
A4:   1

Now, create the frame numbers in column A with the Home/Fill commands that were explained in the first chapter of this book (see Sect. 1.4.1) so that the frame

T. Quirk, *Excel 2010 for Educational and Psychological Statistics:*
A Guide to Solving Practical Problems, DOI 10.1007/978-1-4614-2071-2_2,
© Springer Science+Business Media, LLC 2012

numbers go from 1 to 32, with the number 32 in cell A35. If you need to be reminded about how to do that, here are the steps:

Click on cell A4 to select this cell
Home
Fill (then click on the "down arrow" next to this command and select)
Series (see Fig. 2.1)

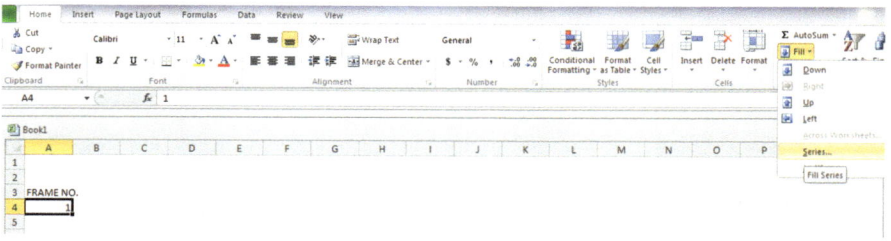

Fig. 2.1 Dialog box for Fill/Series commands

Columns
Step value: 1
Stop value: 32 (see Fig. 2.2)

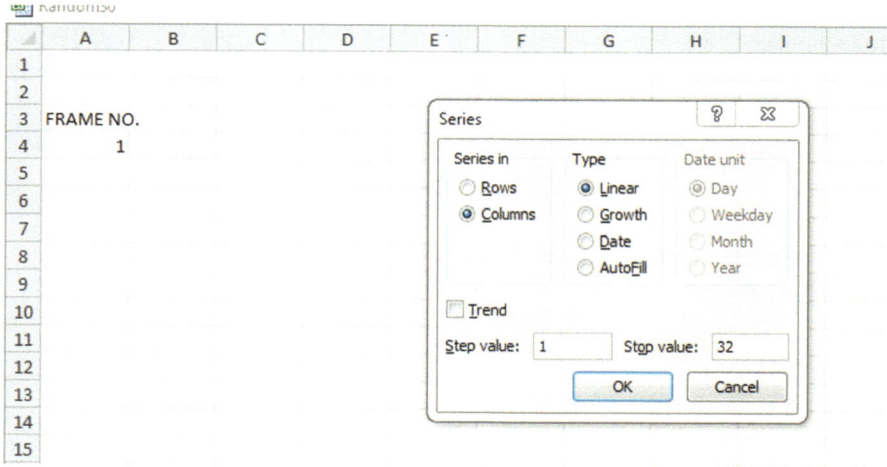

Fig. 2.2 Dialog box for Fill/Series/Columns/Step value/Stop value commands

OK.
Then, save this file as: Random29. You should obtain the result in Fig. 2.3.

Fig. 2.3 Frame numbers
from 1 to 32

FRAME NO.
1
2
3
4
5
6
7
8
9
10
11
12
13
14
15
16
17
18
19
20
21
22
23
24
25
26
27
28
29
30
31
32

Now, create a column next to these frame numbers in this manner.

B3: DUPLICATE FRAME NO.
B4: 1

Next, use the Home/Fill command again so that the 32 frame numbers begin in cell B4 and end in cell B35. Be sure to widen the columns A and B so that all of the information in these columns fits inside the column width. Then, center the information inside both column A and column B on your spreadsheet. You should obtain the information given in Fig. 2.4.

Fig. 2.4 Duplicate frame
numbers from 1 to 32

FRAME NO.	DUPLICATE FRAME NO.
1	1
2	2
3	3
4	4
5	5
6	6
7	7
8	8
9	9
10	10
11	11
12	12
13	13
14	14
15	15
16	16
17	17
18	18
19	19
20	20
21	21
22	22
23	23
24	24
25	25
26	26
27	27
28	28
29	29
30	30
31	31
32	32

Save this file as: Random30

You are probably wondering why you created the same information in both column A and column B of your spreadsheet. This is to make sure that before you sort the frame numbers that you have exactly 32 of them when you finish sorting them into a random sequence of 32 numbers.

Now, let us add a random number to each of the duplicate frame numbers as follows.

2.2 Creating Random Numbers in an Excel Worksheet

C3:   RANDOM NO.

(Then, widen columns A, B, and C so that their labels fit inside the columns; then, center the information in A3:C35).

C4:   =RAND()

Next, hit the enter key to add a random number to cell C4.

Note that you need *both* an open parenthesis *and* a closed parenthesis after =*RAND* (). The RAND command "looks to the left of the cell with the RAND() COMMAND in it" and assigns a random number to that cell.

Now, put the pointer using your mouse in cell C4 and then move the pointer to the bottom right corner of that cell until you see a "plus sign" in that cell. Then, click and drag the pointer down to cell C35 to add a random number to all 32 ID frame numbers (see Fig. 2.5).

FRAME NO.	DUPLICATE FRAME NO.	RANDOM NO.
1	1	0.003582027
2	2	0.782788863
3	3	0.579542097
4	4	0.194826007
5	5	0.180624088
6	6	0.11111051
7	7	0.841557731
8	8	0.43839362
9	9	0.039665164
10	10	0.950687104
11	11	0.701382569
12	12	0.340485937
13	13	0.622362199
14	14	0.823691903
15	15	0.196989157
16	16	0.644490495
17	17	0.818952316
18	18	0.198204913
19	19	0.703849331
20	20	0.412769208
21	21	0.356822645
22	22	0.906835718
23	23	0.770448533
24	24	0.22095226
25	25	0.141239005
26	26	0.035501923
27	27	0.11336016
28	28	0.055916598
29	29	0.368006445
30	30	0.133117174
31	31	0.736884083
32	32	0.702218141

Fig. 2.5 Example of random numbers assigned to the duplicate frame numbers

Then, click on any empty cell to deselect C4:C35 to remove the dark color highlighting these cells.

Save this file as: Random31

Now, let us sort these duplicate frame numbers into a random sequence.

2.3   Sorting Frame Numbers into a Random Sequence

Objective: To sort the duplicate frame numbers into a random sequence

Highlight cells B3:C35 (include the labels at the top of columns B and C).
Data (top of screen).
Sort (click on this word at the top center of your screen; see Fig. 2.6).

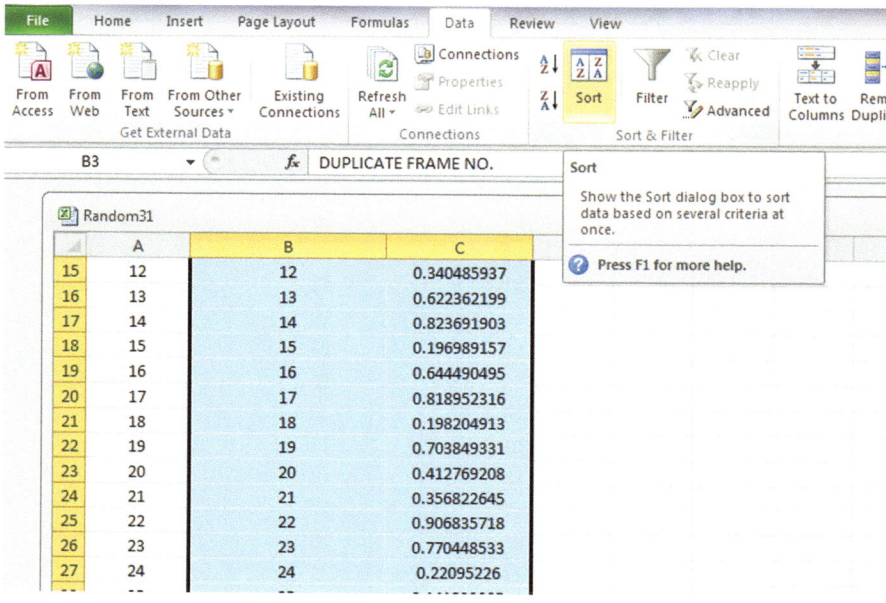

Fig. 2.6 Dialog box for Data/Sort commands

Sort by RANDOM NO. (click on the down arrow)
Smallest to largest (see Fig. 2.7)

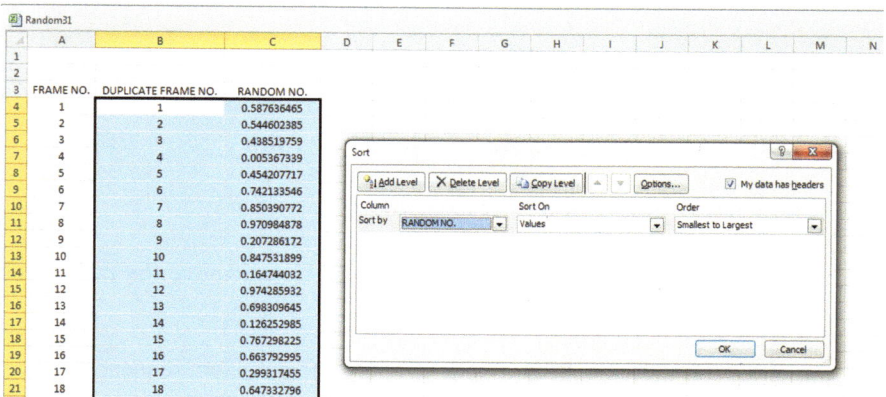

Fig. 2.7 Dialog box for Data/Sort/RANDOM NO./Smallest to Largest commands

OK.

Click on any empty cell to deselect B3:C35.

Save this file as: Random32.

Print this file now.

These steps will produce Fig. 2.8 with the DUPLICATE FRAME NUMBERS sorted into a random order.

FRAME NO.	DUPLICATE FRAME NO.	RANDOM NO.
1	4	0.048919918
2	32	0.715166245
3	19	0.734859307
4	31	0.457996527
5	14	0.837332575
6	25	0.509257077
7	11	0.71473365
8	20	0.297513233
9	29	0.455511676
10	9	0.398344533
11	28	0.074506919
12	17	0.822659637
13	26	0.680326909
14	30	0.586949154
15	3	0.685958841
16	5	0.801702294
17	24	0.042432294
18	27	0.913192281
19	2	0.911877184
20	1	0.000107945
21	22	0.660085877
22	18	0.782601369
23	16	0.222790879
24	13	0.735556843
25	6	0.849257376
26	15	0.232958712
27	10	0.506254928
28	7	0.092256678
29	23	0.990635679
30	21	0.767847232
31	8	0.004497343
32	12	0.748968302

Fig. 2.8 Duplicate frame numbers sorted by random number

Important note: *Because Excel randomly assigns these random numbers, your Excel commands will produce a different sequence of random numbers from everyone else who reads this book!*

Because your objective at the beginning of this chapter was to select randomly 5 of this school's 32 teachers for a personal interview, you now can do that by selecting the *first five ID numbers* in DUPLICATE FRAME NO. column after the sort.

Although your first five random numbers will be different from those we have selected in the random sort that we did in this chapter, we would select these five IDs of teachers to interview using Fig. 2.9.

4, 32, 19, 31, 14

FRAME NO.	DUPLICATE FRAME NO.	RANDOM NO.
1	4	0.048919918
2	32	0.715166245
3	19	0.734859307
4	31	0.457996527
5	14	0.837332575
6	25	0.509257077
7	11	0.71473365
8	20	0.297513233
9	29	0.455511676
10	9	0.398344533
11	28	0.074506919
12	17	0.822659637
13	26	0.680326909
14	30	0.586949154
15	3	0.685958841
16	5	0.801702294
17	24	0.042432294
18	27	0.913192281
19	2	0.911877184
20	1	0.000107945
21	22	0.660085877
22	18	0.782601369
23	16	0.222790879
24	13	0.735556843
25	6	0.849257376
26	15	0.232958712
27	10	0.506254928
28	7	0.092256678
29	23	0.990635679
30	21	0.767847232
31	8	0.004497343
32	12	0.748968302

Fig. 2.9 First five teachers selected randomly

Save this file as: Random33

Remember, your five ID numbers selected after your random sort will be different from the five ID numbers in Fig. 2.9 because Excel assigns a different random number *each time the* $=RAND()$ *command is given.*

Before we leave this chapter, you need to learn how to print a file so that all of the information on that file fits onto a single page without "dribbling over" onto a second or third page.

2.4   Printing an Excel File So That All of the Information Fits onto One Page

> Objective: To print a file so that all of the information fits onto one page

Note that the three practice problems at the end of this chapter require you to sort random numbers when the files contain 63 teachers, 114 counties of the state of Missouri, and 76 key accounts, respectively. These files will be "too big" to fit onto one page when you print them unless you format these files so that they fit onto a single page when you print them.

Let us create a situation where the file does not fit onto one printed page unless you format it first to do that.

Go back to the file you just created, Random 33, and enter the name *Jennifer* into cell A50.

If you printed this file now, the name, *Jennifer*, would be printed onto a second page because it "dribbles over" outside of the page range for this file in its current format.

So, you would need to change the page format so that all of the information, including the name, Jennifer, fits onto just one page when you print this file by using the following steps:

Page layout (top left of the computer screen)

(Notice the "scale to fit" section in the center of your screen; see Fig. 2.10)

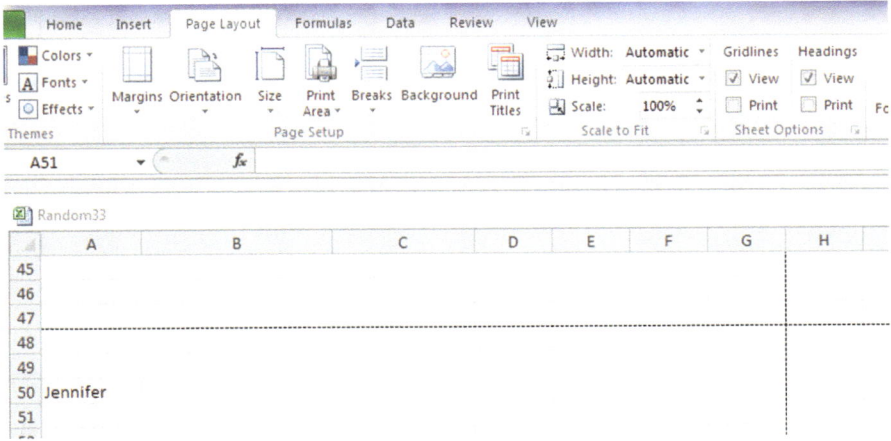

Fig. 2.10 Dialog box for Page Layout/Scale to Fit commands

Hit the down arrow to the right of 100% *once* to reduce the size of the page to 95%.

Now, note that the name, Jennifer, is still on a second page on your screen because her name is below the horizontal dotted line on your screen in Fig. 2.11 (the dotted lines tell you outline dimensions of the file if you printed it now).

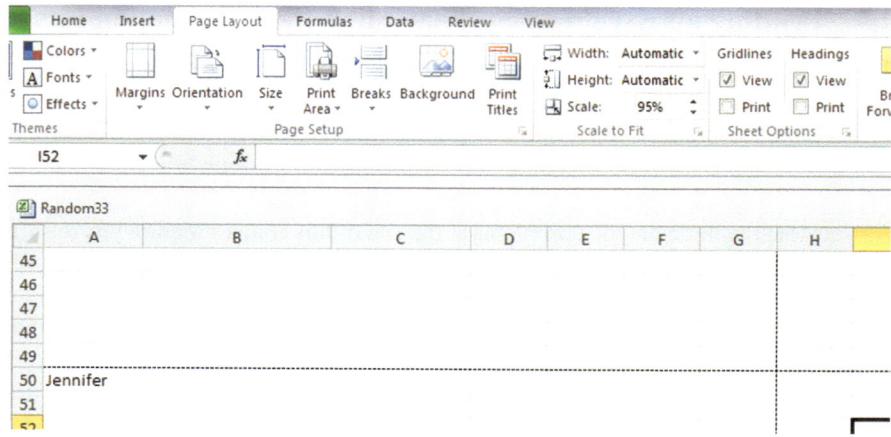

Fig. 2.11 Example of scale reduced to 95% with "Jennifer" to be printed on a second page

So, you need to repeat the "scale change steps" by hitting the down arrow on the right once more to reduce the size of the worksheet to 90% of its normal size.

Notice that the "dotted lines" on your computer screen in Fig. 2.12 are now below Jennifer's name to indicate that all of the information, including her name, is now formatted to fit onto just one page when you print this file.

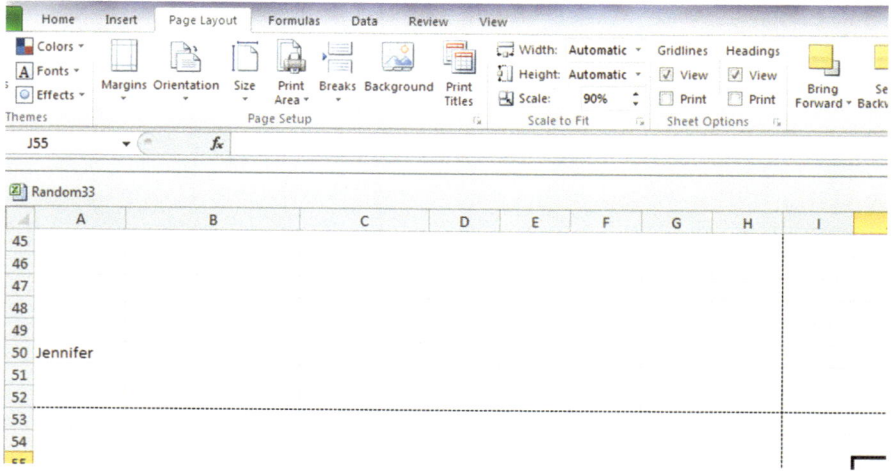

Fig. 2.12 Example of scale reduced to 90% with "Jennifer" to be printed on the first page (note the dotted line below Jennifer on your screen)

Save the file as: Random34
Print the file. Does it all fit onto one page? It should (see Fig. 2.13).

Fig. 2.13 Final spreadsheet
of 90% Scale to Fit

FRAME NO.	DUPLICATE FRAME NO.	RANDOM NO.
1	4	0.882838028
2	32	0.219546518
3	19	0.090551707
4	31	0.372399924
5	14	0.185870948
6	25	0.122282634
7	11	0.76263582
8	20	0.460688523
9	29	0.727708114
10	9	0.201728716
11	28	0.657268728
12	17	0.197213142
13	26	0.073932748
14	30	0.228776006
15	3	0.830419759
16	5	0.696269963
17	24	0.652543027
18	27	0.041499412
19	2	0.108248561
20	1	0.247460826
21	22	0.33124026
22	18	0.433138045
23	16	0.657710193
24	13	0.587800109
25	6	0.06619786
26	15	0.340071703
27	10	0.301135501
28	7	0.198319442
29	23	0.73614172
30	21	0.978573522
31	8	0.250489897
32	12	0.259072675

Jennifer

2.5 End-of-Chapter Practice Problems

1. Suppose that you wanted to do a "customer satisfaction phone survey" of 15 of 63 teachers at a secondary school in St. Louis, Missouri.

   (a) Set up a spreadsheet of frame numbers for these teachers with the heading: FRAME NUMBERS using the Home/Fill commands.

(b) Then, create a separate column to the right of these frame numbers which duplicates these frame numbers with the title: Duplicate frame numbers.

(c) Then, create a separate column to the right of these duplicate frame numbers and use the =RAND() function to assign random numbers to all of the frame numbers in the duplicate frame numbers column and change this column format so that 3 decimal places appear for each random number.

(d) Sort the duplicate frame numbers and random numbers into a random order.

(e) Print the result so that the spreadsheet fits onto one page.

(f) Circle on your printout the ID number of the first 15 teachers that you would call in your phone survey.

(g) Save the file as RAND9.

Important note: *Note that everyone who does this problem will generate a different random order of teacher ID numbers since Excel assigns a different random number each time the RAND() command is used. For this reason, the answer to this problem given in this Excel Guide will have a completely different sequence of random numbers from the random sequence that you generate. This is normal and what is to be expected.*

2. Suppose that you wanted to do a random sample of 10 of the 114 counties in the state of Missouri as requested by a political pollster who wants to select registered voters by county in Missouri for a phone survey of their voting preferences in the next election. You know that there are 114 counties in Missouri because you have accessed the Web site for the US census (U.S. Census Bureau 2000). For your information, the United States has a total of 3,140 counties in its 50 states (U.S. Census Bureau 2000).

(a) Set up a spreadsheet of frame numbers for these counties with the heading: FRAME NO.

(b) Then, create a separate column to the right of these frame numbers which duplicates these frame numbers with the title: Duplicate frame no.

(c) Then, create a separate column to the right of these duplicate frame numbers entitled "random number" and use the =RAND() function to assign random numbers to all of the frame numbers in the duplicate frame numbers column. Then, change this column format so that 3 decimal places appear for each random number.

(d) Sort the duplicate frame numbers and random numbers into a random order.

(e) Print the result so that the spreadsheet fits onto one page.

(f) Circle on your printout the ID number of the first 10 counties that the pollster would call in his phone survey.

(g) Save the file as RANDOM6.

3. Suppose that a sales department at a company wants to do a "customer satisfaction survey" of 20 of this company's 76 "key accounts." Suppose, further, that the sales vice-president has defined a key account as a customer who purchased at least $30,000 worth of merchandise from this company in the past 90 days.

(a) Set up a spreadsheet of frame numbers for these customers with the heading: FRAME NUMBERS.
(b) Then, create a separate column to the right of these frame numbers which duplicates these frame numbers with the title: Duplicate frame numbers.
(c) Then, create a separate column to the right of these duplicate frame numbers entitled "random number" and use the =RAND() function to assign random numbers to all of the frame numbers in the duplicate frame numbers column. Then, change this column format so that 3 decimal places appear for each random number.
(d) Sort the duplicate frame numbers and random numbers into a random order.
(e) Print the result so that the spreadsheet fits onto one page.
(f) Circle on your printout the ID number of the first 20 customers that the sales vice-president would call for his phone survey.
(g) Save the file as RAND5.

Reference

U.S. Census Bureau Census 2000 PHC-T-4. Ranking tables for counties 1990 and 2000. Retrieved from http://www.census.gov/population/www/cen2000/briefs/phc-t4/tables/tab01.pdf

Chapter 3
Confidence Interval About the Mean Using the TINV Function and Hypothesis Testing

This chapter focuses on two ideas: (1) finding the 95% confidence interval about the mean and (2) hypothesis testing.

Let us talk about the confidence interval first.

3.1 Confidence Interval About the Mean

In statistics, we are always interested in *estimating the population mean.* How do we do that?

3.1.1 How to Estimate the Population Mean

Objective: To estimate the population mean, μ

Remember that the population mean is the average of all of the people in the target population. For example, if we were interested in how well adults ages 25–44 liked a new flavor of Ben & Jerry's ice cream, we could never ask this question to all of the people in the USA who were in that age group. Such a research study would take way too much time to complete, and the cost of doing that study would be prohibitive.

So, instead of testing *everyone* in the population, we take a sample of people in the population and use the results of this sample to estimate the mean of the entire population. This saves both time and money. When we use the results of a sample to estimate the population mean, this is called "*inferential statistics*" because we are inferring the population mean from the sample mean.

When we study a sample of people in educational or psychological research, we know the size of our sample (n), the mean of our sample ($\overline{X}$), and the standard deviation of our sample (STDEV). We use these figures to estimate the population mean with a test called the "confidence interval about the mean."

T. Quirk, *Excel 2010 for Educational and Psychological Statistics:* 35
A Guide to Solving Practical Problems, DOI 10.1007/978-1-4614-2071-2_3,
© Springer Science+Business Media, LLC 2012

3.1.2 *Estimating the Lower Limit and the Upper Limit of the 95% Confidence Interval About the Mean*

The theoretical background of this test is beyond the scope of this book, and you can learn more about this test from studying any good statistics textbook (e.g., Levine 2011), but the basic ideas are as follows:

We assume that the population mean is somewhere in an interval which has a "lower limit" and an "upper limit" to it. We also assume in this book that we want to be "95% confident" that the population mean is inside this interval somewhere. So, we intend to make the following type of statement:

"We are 95% confident that the population mean in miles per gallon (mpg) for the Chevy Impala automobile is between 26.92 miles per gallon and 29.42 miles per gallon."

If we want to create a billboard for this car that claims that this car gets 28 miles per gallon (mpg), we can do that because 28 is *inside the 95% confidence interval* in our research study in the above example. We do not know exactly what the population mean is, only that it is somewhere between 26.92 and 29.42 mpg, and 28 is inside this interval.

But we are only 95% confident that the population mean is inside this interval, and 5% of the time we will be wrong in assuming that the population mean is 28 mpg.

But, for our purposes in educational and psychological research, we are happy to be 95% confident that our assumption is accurate. We should also point out that 95% is an arbitrary level of confidence for our results. We could choose to be 80% confident or 90% confident or even 99% confident in our results if we wanted to do that. But, in this book, *we will always assume that we want to be 95% confident of our results.* That way, you will not have to guess on how confident you want to be in any of the problems in this book. We will always want to be 95% confident of our results in this book.

So how do we find the 95% confidence interval about the mean for our data?

In words, we will find this interval this way:

"Take the sample mean ($\overline{X}$) *and add to it* 1.96 times the standard error of the mean (s.e.) to get the upper limit of the confidence interval. Then, take the sample mean *and subtract from it* 1.96 times the standard error of the mean to get the lower limit of the confidence interval."

You will remember (see Sect. 1.3) that the standard error of the mean (s.e.) is found by dividing the standard deviation of our sample (STDEV) by the square root of our sample size (n).

In mathematical terms, the formula for the 95% confidence interval about the mean is

$$\overline{X} \pm 1.96 \text{ s.e.} \tag{3.1}$$

Note that the "±"sign stands for "plus or minus," and this means that you first add 1.96 times the s.e. to the mean to get the upper limit of the confidence interval, and then subtract 1.96 times the s.e. from the mean to get the lower limit of the confidence interval. Also, the symbol 1.96 s.e. means that you multiply 1.96 times the standard error of the mean to get this part of the formula for the confidence interval.

Note: We will explain shortly where the number 1.96 came from.

Let us try a simple example to illustrate this formula.

3.1.3 Estimating the Confidence Interval for the Chevy Impala in Miles per Gallon

Let us suppose that you asked owners of the Chevy Impala to keep track of their mileage and the number of gallons used for two tanks of gas. Let us suppose that 49 owners did this, and that they average 27.83 miles per gallon (mpg) with a standard deviation of 3.01 mpg. The standard error (s.e.) would be 3.01 divided by the square root of 49 (i.e., 7) which gives an s.e. equal to 0.43.

The 95% confidence interval for these data would be:

$$27.83 \pm 1.96(0.43).$$

The *upper limit of this confidence interval* uses the plus sign of the ± sign in the formula. Therefore, the upper limit would be:

$$27.83 + 1.96(0.43) = 27.83 + 0.84 = 28.67 \, \text{mpg}.$$

Similarly, *the lower limit of this confidence interval* uses the minus sign of the ± sign in the formula. Therefore, the lower limit would be:

$$27.83 - 1.96(0.43) = 27.83 - 0.84 = 26.99 \, \text{mpg}.$$

The result of our research study would, therefore, be the following:

"We are 95% confident that the population mean for the Chevy Impala is somewhere between 26.99 mpg and 28.67 mpg."

If we were planning to create a billboard that claimed that this car got 28 mpg, we would be able to do that based on our data since 28 is inside of this 95% confidence interval for the population mean.

You are probably asking yourself: "Where did that 1.96 in the formula come from?"

3.1.4   Where Did the Number "1.96" Come From?

A detailed mathematical answer to that question is beyond the scope of this book, but here is the basic idea.

We make an assumption that the data in the population are "normally distributed" in the sense that the population data would take the shape of a "normal curve" if we could test all of the people in the population. The normal curve looks like the outline of the Liberty Bell that sits in front of Independence Hall in Philadelphia, Pennsylvania. The normal curve is "symmetric" in the sense that if we cut it down the middle and folded it over to one side, the half that we folded over would fit perfectly onto the half on the other side.

A discussion of integral calculus is beyond the scope of this book, but essentially, we want to find the lower limit and the upper limit of the population data in the normal curve so that 95% of the area under this curve is between these two limits. *If we have more than 40 people in our research study*, the value of these limits is plus or minus 1.96 times the standard error of the mean (s.e.) of our sample. The number 1.96 times the s.e. of our sample gives us the upper limit and the lower limit of our confidence interval. If you want to learn more about this idea, you can consult a good statistics book (e.g., Salkind 2010).

The number 1.96 would change if we wanted to be confident of our results at a different level from 95% as long as we have more than 40 people in our research study.

For example

1. If we wanted to be 80% confident of our results, this number would be 1.282.
2. If we wanted to be 90% confident of our results, this number would be 1.645.
3. If we wanted to be 99% confident of our results, this number would be 2.576.

But since we always want to be 95% confident of our results in this book, we will always use 1.96 in this book whenever we have more than 40 people in our research study.

By now, you are probably asking yourself: "Is this number in the confidence interval about the mean always 1.96?" The answer is "No!," and we will explain why this is true now.

3.1.5   Finding the Value for t in the Confidence Interval Formula

Objective: To find the value for t in the confidence interval formula

The correct formula for the confidence interval about the mean for different sample sizes is the following:

$$\overline{X} \pm t \text{ s.e.} \tag{3.2}$$

To use this formula, you find the sample mean, $\overline{X}$, *and add to it the value of t times the s.e. to get the upper limit* of this 95% confidence interval. Also, you take the sample mean, $\overline{X}$, and *subtract from it the value of t times the s.e. to get the lower limit* of this 95% confidence interval. And you find the value of t in the table given in Appendix E of this book in the following way:

Objective: To find the value of t in the t-table in Appendix E

Before we get into an explanation of what is meant by "the value of t," let us give you practice in finding the value of t by using the t-table in Appendix E.

Keep your finger on Appendix E as we explain how you need to "read" that table.

Since the test in this chapter is called the "confidence interval about the mean test," you will use the first column on the left in Appendix E to find the critical value of t for your research study (note that this column is headed "sample size n").

To find the value of t, you go down this first column until you find the sample size in your research study and then you go to the right and read the value of t for that sample size in the "critical t column" of the table (note that this column is the column that you would use for the 95% confidence interval about the mean).

For example, if you have 14 people in your research study, the value of t is 2.160.

If you have 26 people in your research study, the value of t is 2.060.

If you have more than 40 people in your research study, the value of t is always 1.96.

Note that the "critical t column" in Appendix E represents the value of t that you need to use to obtain to be 95% confident of your results as "significant" results.

Throughout this book, we are assuming that you want to be 95% confident in the results of your statistical tests. Therefore, the value for t in the t-table in Appendix E tells you which value you should use for t when you use the formula for the 95% confidence interval about the mean.

Now that you know how to find the value of t in the formula for the confidence interval about the mean, let us explore how you find this confidence interval using Excel.

3.1.6   Using Excel's TINV Function to Find the Confidence Interval About the Mean

---

Objective: To use the TINV function in Excel to find the confidence interval about the mean

---

When you use Excel, the formulas for finding the confidence interval are the following:

$$\text{Lower limit:} = \overline{X} - \text{TINV}(1 - 0.95, n - 1) * \text{s.e.}$$
$$\text{(no spaces between these symbols)} \tag{3.3}$$

$$\text{Upper limit: } = \overline{X} + \text{TINV}(1 - 0.95, n - 1) * \text{s.e.}$$
$$\text{(no spaces between these symbols)} \tag{3.4}$$

Note that the "*" symbol in this formula tells Excel to use the multiplication step in the formula, and it stands for "times" in the way we talk about multiplication.

You will recall from Chap. 1 that n stands for the sample size, and so, $n - 1$ stands for the sample size minus one.

You will also recall from Chap. 1 that the standard error of the mean, s.e., equals the STDEV divided by the square root of the sample size, n (see Sect. 1.3).

Let us try a sample problem using Excel to find the 95% confidence interval about the mean for a problem.

Let us suppose that General Motors wanted to claim that its Chevy Impala achieves 28 miles per gallon (mpg) on the highway. Let us call 28 mpg the "reference value" for this car.

Suppose that you work for Ford Motor Co. and that you want to check this claim to see if it holds up based on some research evidence. You decide to collect some data and use a two-side 95% confidence interval about the mean to test your results:

3.1.7 Using Excel to Find the 95% Confidence Interval for a Car's Miles per Gallon Claim

Objective: To analyze the data using a two-side 95% confidence interval about the mean

You select a sample of new car owners for this car, and they agree to keep track of their mileage for two tanks of gas and to record the average miles per gallon they achieve on these two tanks of gas. Your research study produces the results given in Fig. 3.1:

Create a spreadsheet with these data and use Excel to find the sample size (n), the mean, the standard deviation (STDEV), and the standard error of the mean (s.e.) for these data using the following cell references:

A3:   Chevy Impala
A5:   Miles per gallon
A6:   30.9
Enter the other mpg data in cells A7:A30

Chevy Impala

Miles per gallon
     30.9
     24.5
     31.2
     28.7
     35.1
     29.0
     28.8
     23.1
     31.0
     30.2
     28.4
     29.3
     24.2
     27.0
     26.7
     31.0
     23.5
     29.4
     26.3
     27.5
     28.2
     28.4
     29.1
     21.9
     30.9

Fig. 3.1 Worksheet data for Chevy Impala (practical example)

Now, highlight cells A6:A30 and format these numbers in number format (one decimal place). Center these numbers in column A. Then, widen columns A and B by making both of them twice as wide as the original width of column A. Then, widen column C so that it is three times as wide as the original width of column A so that your table looks more professional.

C7:   *n*
C10:   Mean
C13:   STDEV
C16:   s.e.
C19:   95% confidence interval
D21:   Lower limit:
D23:   Upper limit: (see Fig. 3.2)

Chevy Impala

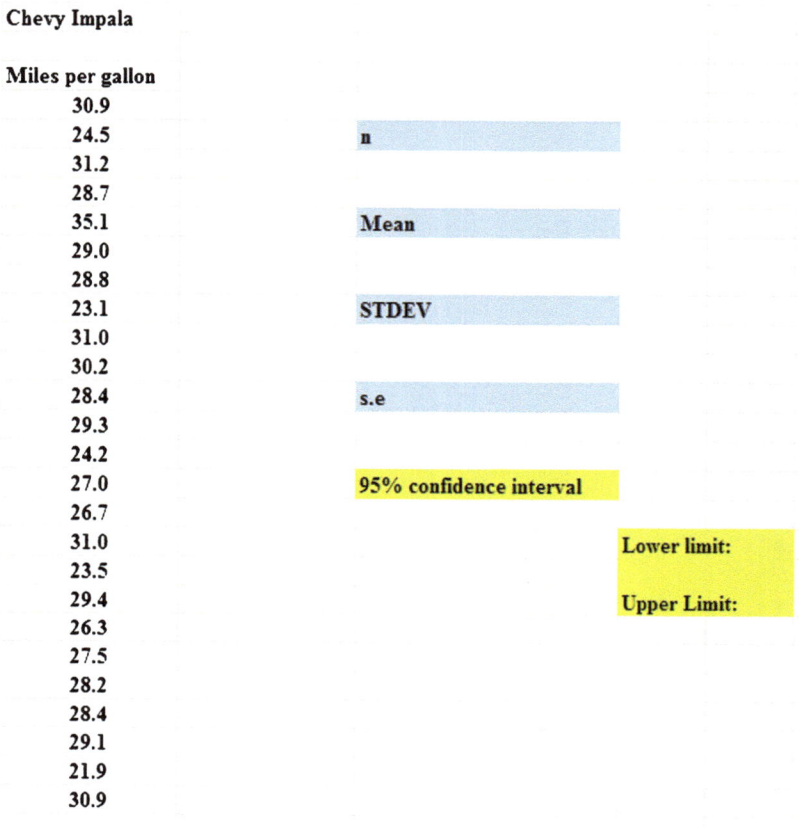

Miles per gallon
      30.9
      24.5   n
      31.2
      28.7
      35.1   Mean
      29.0
      28.8
      23.1   STDEV
      31.0
      30.2
      28.4   s.e
      29.3
      24.2
      27.0   95% confidence interval
      26.7
      31.0   Lower limit:
      23.5
      29.4   Upper Limit:
      26.3
      27.5
      28.2
      28.4
      29.1
      21.9
      30.9

Fig. 3.2 Example of Chevy Impala format for the confidence interval about the mean labels

B26:   Draw a picture below this confidence interval
B28:   26.92
B29:   lower
B30:   limit
C28:   '........... 28 .... 28.17 .... (note that you need to begin cell C28 with
         a *single quotation mark* (') to tell Excel that this is a *label*, and not a
         number)
E28:   '29.42 (note the single quotation mark)
C29:   ref. Mean
C30:   value
E29:   upper
E30:   limit
B33:   Conclusion:

Now, align the labels underneath the picture of the confidence interval so that they will look like Fig. 3.3.

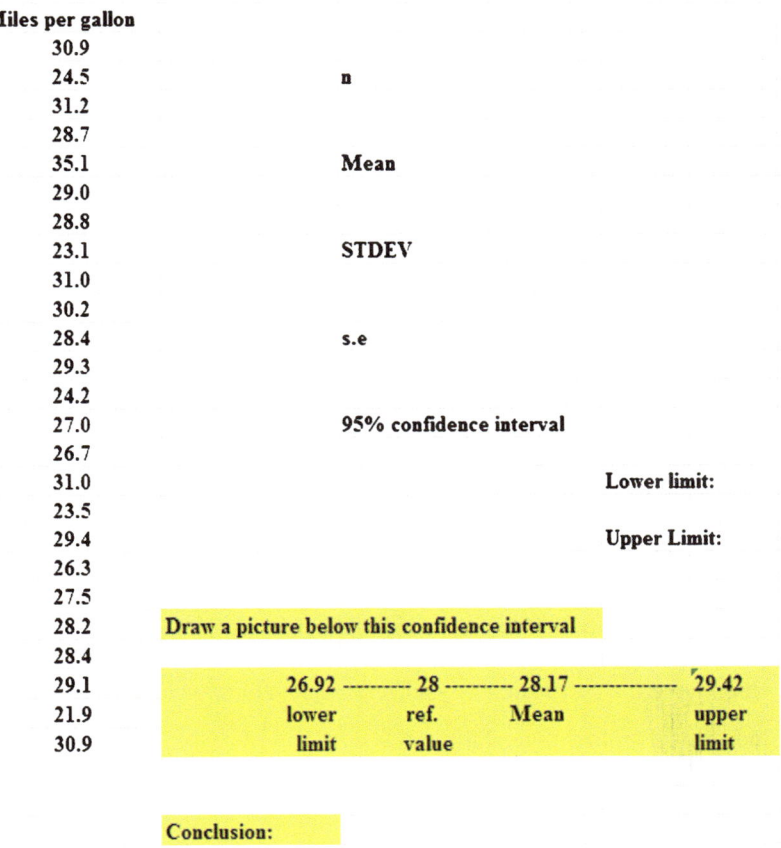

Chevy Impala

Miles per gallon

30.9	
24.5	n
31.2	
28.7	
35.1	Mean
29.0	
28.8	
23.1	STDEV
31.0	
30.2	
28.4	s.e
29.3	
24.2	
27.0	95% confidence interval
26.7	
31.0	Lower limit:
23.5	
29.4	Upper Limit:
26.3	
27.5	
28.2	Draw a picture below this confidence interval
28.4	
29.1	26.92 ---------- 28 ---------- 28.17 -------------- 29.42
21.9	lower ref. Mean upper
30.9	limit value limit

Conclusion:

Fig. 3.3 Example of drawing a picture of a confidence interval about the mean result

Next, name the range of data from A6:A30 as: miles

D7:   Use Excel to find the sample size
D10: Use Excel to find the mean
D13: Use Excel to find the STDEV
D16: Use Excel to find the s.e.

Now, you need to find the lower limit and the upper limit of the 95% confidence interval for this study.

We will use Excel's TINV function to do this. We will assume that you want to be 95% confident of your results.

F21: =D10−TINV(1−.95,24)*D16

Note that this TINV formula uses 24 since 24 is one less than the sample size of 25 (i.e., 24 is $n - 1$). Note that D10 is the mean, while D16 is the standard error of the mean. The above formula gives the *lower limit of the confidence interval, 26.92.*

F23: =D10+TINV(1−.95,24)*D16

The above formula gives the *upper limit of the confidence interval, 29.42.*

Now, use number format (two decimal places) in your Excel spreadsheet for the mean, standard deviation, standard error of the mean, and for both the lower limit and the upper limit of your confidence interval. If you printed this spreadsheet now, the lower limit of the confidence interval (26.92) and the upper limit of the confidence interval (29.42) would "dribble over" onto a second printed page because the information on the spreadsheet is too large to fit onto one page in its present format.

So, you need to use Excel's "Scale to Fit" commands that we discussed in Chap. 2 (see Sect. 2.4) to reduce the size of the spreadsheet to 95% of its current size using the Page Layout/Scale to Fit function. Do that now, and notice that the dotted line to the right of 26.92 and 29.42 indicates that these numbers would now fit onto one page when the spreadsheet is printed out (see Fig. 3.4).

Note that you have drawn a picture of the 95% confidence interval beneath cell B26, including the lower limit, the upper limit, the mean, and the reference value of 28 mpg, given in the claim that the company wants to make about the car's miles per gallon performance.

Now, let us write the conclusion to your research study on your spreadsheet:

C33:   Since the reference value of 28 is inside
C34:   the confidence interval, we accept that
C35:   the Chevy Impala does get 28 mpg.

Your research study accepted the claim that the Chevy Impala did get 28 miles per gallon on the highway. The average miles per gallon in your study was 28.17 (see Fig. 3.5).

Save your resulting spreadsheet as: *CHEVY7.*

3.2 Hypothesis Testing

One of the important activities of researchers, whether they are in educational or psychological research, marketing research, or in any of the social sciences, is that they attempt to "check" their assumptions about the world by testing these assumptions in the form of hypotheses.

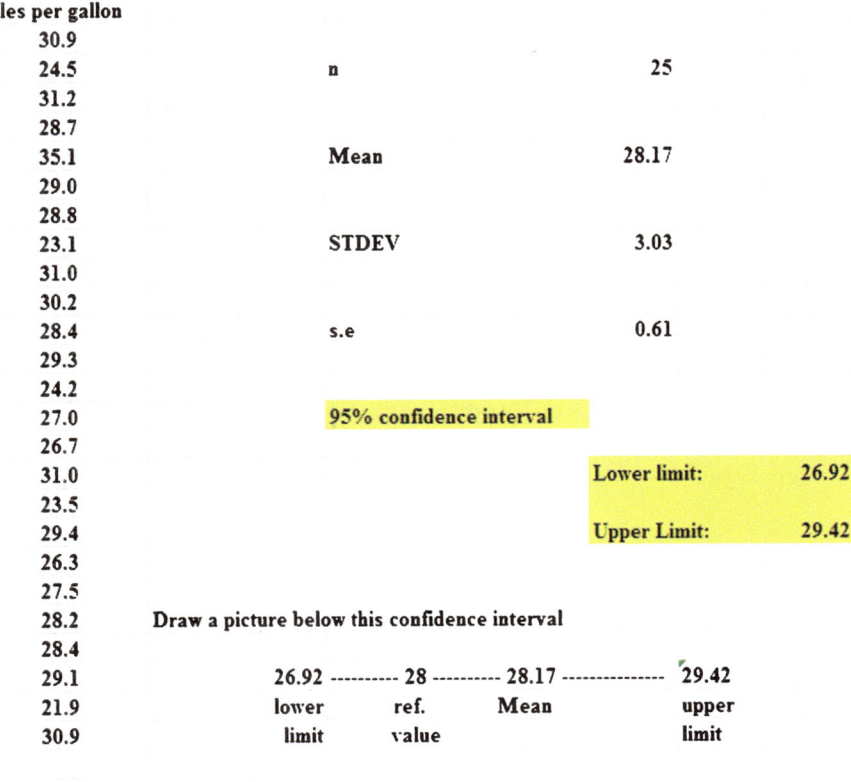

Chevy Impala

Miles per gallon
30.9		
24.5	n	25
31.2		
28.7		
35.1	Mean	28.17
29.0		
28.8		
23.1	STDEV	3.03
31.0		
30.2		
28.4	s.e	0.61
29.3		
24.2		
27.0	95% confidence interval	
26.7		
31.0	Lower limit:	26.92
23.5		
29.4	Upper Limit:	29.42
26.3		
27.5		
28.2	Draw a picture below this confidence interval	
28.4		
29.1	26.92 --------- 28 --------- 28.17 -------------- 29.42	
21.9	lower ref. Mean upper	
30.9	limit value limit	

Conclusion:

Fig. 3.4 Result of using the TINV function to find the confidence interval about the mean

A typical hypothesis is in the form: *"If x, then y."*
Some examples would be:

1. If we use this new method of teaching reading in fourth grade classes, the Iowa Test of Basic Skills (ITBS) grade equivalent (GE) reading scores will improve by 5%.
2. If we raise our price by 5%, then our sales dollars for this product will decrease by 4% but our profit for this product will increase by 3%.
3. If we use this new method of teaching mathematics to ninth graders in algebra, then our math achievement scores will go up by 10%.

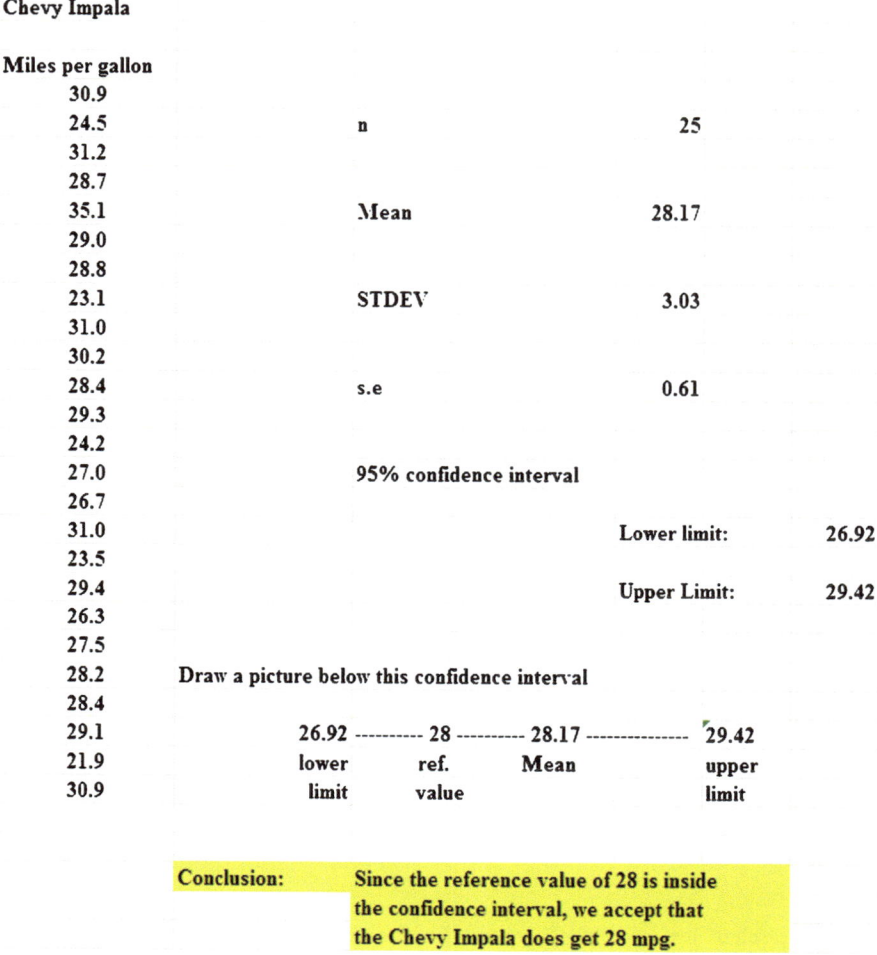

Chevy Impala

Miles per gallon

30.9		
24.5	n	25
31.2		
28.7		
35.1	Mean	28.17
29.0		
28.8		
23.1	STDEV	3.03
31.0		
30.2		
28.4	s.e	0.61
29.3		
24.2		
27.0	95% confidence interval	
26.7		
31.0	Lower limit:	26.92
23.5		
29.4	Upper Limit:	29.42
26.3		
27.5		
28.2	Draw a picture below this confidence interval	
28.4		
29.1	26.92 ---------- 28 --------- 28.17 -------------- 29.42	
21.9	lower ref. Mean upper	
30.9	limit value limit	

Conclusion: Since the reference value of 28 is inside
the confidence interval, we accept that
the Chevy Impala does get 28 mpg.

Fig. 3.5 Final spreadsheet for the Chevy Impala confidence interval about the mean

4. If we change the format for teaching Introductory Psychology to our undergraduates, then their final exam scores will increase by 10%.

A hypothesis, then, to a social science researcher is a "guess" about what we think is true in the real world. We can test these guesses using statistical formulas to see if our predictions come true in the real world.

So, in order to perform these statistical tests, we must first state our hypotheses so that we can test our results against our hypotheses to see if our hypotheses match reality.

So, how do we generate hypotheses in education or psychology?

3.2.1   Hypotheses Always Refer to the Population of People or Events That You Are Studying

The first step is to understand that our hypotheses always refer to the *population* of people under study.

For example, if we are interested in studying 18–24 year-olds in St. Louis as our target market, and we select a sample of people in this age group in St. Louis, depending on how we select our sample, we are hoping that our results of this study are useful in generalizing our findings to *all* 18–24 year-olds in St. Louis, and not just to the particular people in our sample.

The entire group of 18–24 year-olds in St. Louis would be the *population* that we are interested in studying, while the particular group of people in our study is called the *sample* from this population.

Since our sample sizes typically contain only a few people, we are interested in the results of our sample *only insofar as the results of our sample can be "generalized" to the population in which we are really interested.*

That is why our hypotheses always refer to the population and never to the sample of people in our study.

You will recall from Chap. 1 that we used the symbol $\overline{X}$ to refer to the mean of the sample we use in our research study (see Sect. 1.1).

We will use the symbol μ (the Greek letter "mu") to refer to the *population mean*.

In testing our hypotheses, we are trying to decide which one of two competing hypotheses *about the population mean* we should accept given our data set.

3.2.2   The Null Hypothesis and the Research (Alternative) Hypothesis

These two hypotheses are called the *null hypothesis* and the *research hypothesis*.

Statistics textbooks typically refer to the *null hypothesis* with the notation H_0.

The *research hypothesis* is typically referred to with the notation H_1, and it is sometimes called the *alternative hypothesis*.

Let us explain first what is meant by the null hypothesis and the research hypothesis:

1. *The null hypothesis is what we accept as true unless we have compelling evidence that it is not true.*
2. *The research hypothesis is what we accept as true whenever we reject the null hypothesis as true.*

This is similar to our legal system in America where we assume that a supposed criminal is innocent until he or she is proven guilty in the eyes of a jury. Our null hypothesis is that this defendant is innocent, while the research hypothesis is that he or she is guilty.

In the great state of Missouri, every license plate has the state slogan "Show me." This means that people in Missouri think of themselves as not gullible enough to accept everything that someone says as true unless that person's actions indicate the truth of his or her claim. In other words, people in Missouri believe strongly that a person's actions speak much louder than that person's words.

Since both the null hypothesis and the research hypothesis cannot both be true, the task of hypothesis testing using statistical formulas is to decide which one you will accept as true and which one you will reject as true.

Sometimes in business research, a series of rating scales is used to measure people's attitudes toward a company, toward one of its products, or toward their intention to buy that company's products. These rating scales are typically five-point, seven-point, or ten-point scales, although other scale values are often used as well.

3.2.2.1 Determining the Null Hypothesis and the Research Hypothesis When Rating Scales Are Used

Here is a typical example of a seven-point scale in education for parents of eighth-grade pupils at the end of a school year (see Fig. 3.6):

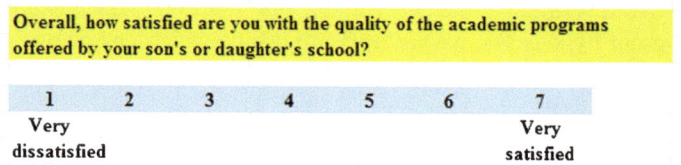

Fig. 3.6 Example of a rating scale item for parents of eighth graders (practical example)

So, how do we decide what to use as the null hypothesis and the research hypothesis whenever rating scales are used?

Objective: To decide on the null hypothesis and the research hypothesis when-ever rating scales are used

In order to make this determination, we will use a simple rule:

Rule: *Whenever rating scales are used, we will use the "middle" of the scale as the null hypothesis and the research hypothesis.*

In the above example, since 4 is the number in the middle of the scale (i.e., three numbers are below it, and three numbers are above it), our hypotheses become:

Null hypothesis: $\mu = 4$
Research hypothesis: $\mu \neq 4$

In the above rating scale example, if the result of our statistical test for this one attitude scale item indicates that our population mean is "close to 4," we say that we accept the null hypothesis that the parents of eighth-grade pupils were neither satisfied nor dissatisfied with the quality of the academic programs offered by their son's or daughter's school.

In the above example, if the result of our statistical test indicates that the population mean is significantly different from 4, we reject the null hypothesis and accept the research hypothesis *by stating either that*:

"*Parents of 8th grade pupils were significantly satisfied with the quality of the academic programs offered by their son's or daughter's school*" (this is true whenever our sample mean is significantly greater than our expected population mean of 4).

or

"*Parents of 8th grade pupils were significantly dissatisfied with the quality of the academic programs offered by their son's or daughter's school*" (this is accepted as true whenever our sample mean is significantly less than our expected population mean of 4).

Both of these conclusions cannot be true. We accept one of the hypotheses as "true" based on the data set in our research study, and the other one as "not true" based on our data set.

The job of the educational and psychological researcher, then, is to decide which of these two hypotheses, the null hypothesis or the research hypothesis, he or she will accept as true given the data set in the research study.

Let us try some examples of rating scales so that you can practice figuring out what the null hypothesis and the research hypothesis are for each rating scale.

In the spaces in Fig. 3.7, write in the null hypothesis and the research hypothesis for the rating scales:

How did you do?

Here are the answers to these three questions:

1. The null hypothesis is $\mu = 3$, and the research hypothesis is $\mu \neq 3$ on this five-point scale (i.e. the "middle" of the scale is 3).
2. The null hypothesis is $\mu = 4$, and the research hypothesis is $\mu \neq 4$ on this seven-point scale (i.e., the "middle" of the scale is 4).
3. The null hypothesis is $\mu = 5.5$, and the research hypothesis is $\mu \neq 5.5$ on this ten-point scale (i.e., the "middle" of the scale is 5.5 since there are 5 numbers below 5.5 and 5 numbers above 5.5).

As another example, Holiday Inn Express in its Stay Smart Experience Survey uses four-point scales where:

1 = Not so good
2 = Average
3 = Very good
4 = Great

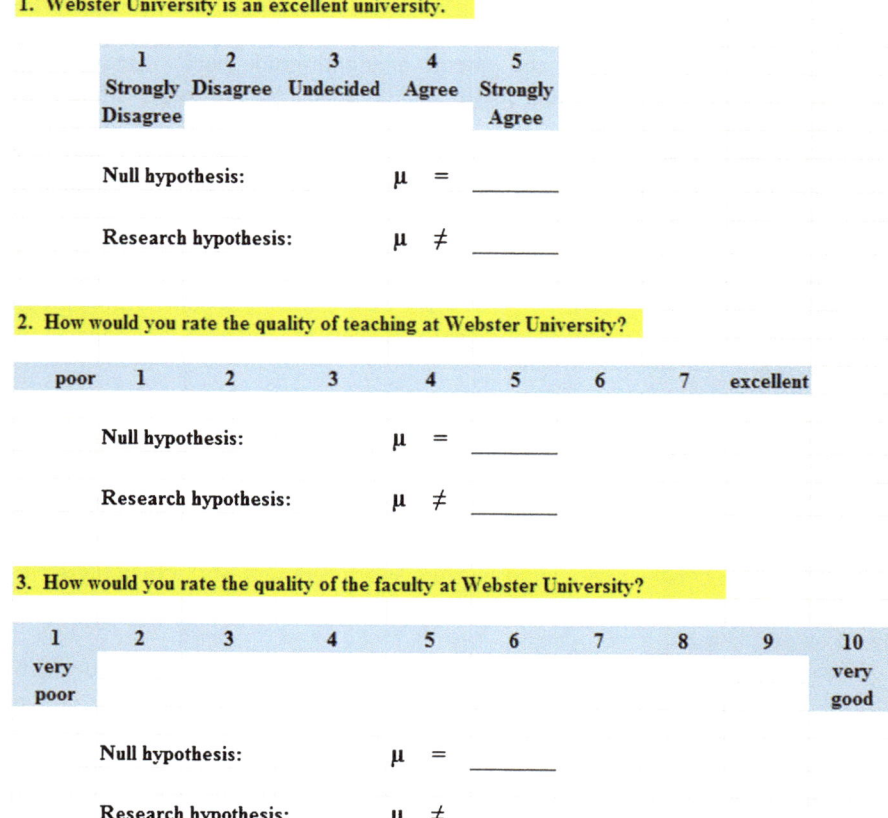

1. Webster University is an excellent university.

1	2	3	4	5
Strongly Disagree	Disagree	Undecided	Agree	Strongly Agree

Null hypothesis: μ = _____

Research hypothesis: μ ≠ _____

2. How would you rate the quality of teaching at Webster University?

poor	1	2	3	4	5	6	7	excellent

Null hypothesis: μ = _____

Research hypothesis: μ ≠ _____

3. How would you rate the quality of the faculty at Webster University?

1	2	3	4	5	6	7	8	9	10
very poor									very good

Null hypothesis: μ = _____

Research hypothesis: μ ≠ _____

Fig. 3.7 Examples of rating scales for determining the null hypothesis and the research hypothesis

On this scale, the null hypothesis is $\mu = 2.5$, and the research hypothesis is $\mu \neq 2.5$ because there are two numbers below 2.5 and two numbers above 2.5 on that rating scale.

Now, let us discuss the seven steps of hypothesis testing for using the confidence interval about the mean.

3.2.3   The Seven Steps for Hypothesis Testing Using the Confidence Interval About the Mean

---

Objective: To learn the seven steps of hypothesis testing using the confidence interval about the mean

---

There are seven basic steps of hypothesis testing for this statistical test.

3.2.3.1 Step 1: State the Null Hypothesis and the Research Hypothesis

If you are using numerical scales in your survey, you need to remember that these hypotheses refer to the "middle" of the numerical scale. For example, if you are using seven-point scales with $1 =$ poor and $7 =$ excellent, these hypotheses would refer to the middle of these scales and would be:

Null hypothesis H_0: $\quad \mu = 4$
Research hypothesis H_1: $\quad \mu \neq 4$

3.2.3.2 Step 2: Select the Appropriate Statistical Test

In this chapter, we are studying the confidence interval about the mean, and so we will select that test.

3.2.3.3 Step 3: Calculate the Formula for the Statistical Test

You will recall (see Sect. 3.1.5) that the formula for the confidence interval about the mean is

$$\overline{X} \pm t \text{ s.e.} \tag{3.2}$$

We discussed the procedure for computing this formula for the confidence interval about the mean using Excel earlier in this chapter, and the steps involved in using that formula are:

1. Use Excel's =COUNT function to find the sample size.
2. Use Excel's =AVERAGE function to find the sample mean, $\overline{X}$.
3. Use Excel's =STDEV function to find the standard deviation, STDEV.
4. Find the standard error of the mean (s.e.) by dividing the standard deviation (STDEV) by the square root of the sample size, n.
5. Use Excel's TINV function to find the lower limit of the confidence interval.
6. Use Excel's TINV function to find the upper limit of the confidence interval.

3.2.3.4 Step 4: Draw a Picture of the Confidence Interval About the Mean, Including the Mean, the Lower Limit of the Interval, the Upper Limit of the Interval, and the Reference Value Given in the Null Hypothesis, H_0

(We will explain step 4 later in the Chapter)

3.2.3.5 Step 5: Decide on a Decision Rule

(a) *If the reference value is inside the confidence interval, accept the null hypothesis, H_0.*

(b) *If the reference value is outside the confidence interval, reject the null hypothesis, H_0, and accept the research hypothesis, H_1.*

3.2.3.6 Step 6: State the Result of Your Statistical Test

There are two possible results when you use the confidence interval about the mean, and only one of them can be accepted as "true." So your result would be one of the following:

Either: Since the reference value is inside the confidence interval, *we accept the null hypothesis, H_0.*

Or: Since the reference value is outside the confidence interval, *we reject the null hypothesis, H_0, and accept the research hypothesis, H_1.*

3.2.3.7 Step 7: State the Conclusion of Your Statistical Test in Plain English!

In practice, this is more difficult than it sounds because you are trying to summarize the result of your statistical test in simple English that is both concise and accurate so that someone who has never had a statistics course (such as your boss, perhaps) can understand the conclusion of your test. This is a difficult task, and we will give you a lot of practice doing this last and most important step throughout this book.

> Objective: To write the conclusion of the confidence interval about the mean test

Let us set some basic rules for stating the conclusion of a hypothesis test.

Rule #1: *Whenever you reject H_0 and accept H_1, you must use the word "significantly" in the conclusion to alert the reader that this test found an important result.*

Rule #2: *Create an outline in words of the "key terms" you want to include in your conclusion so that you do not forget to include some of them.*

Rule #3: *Write the conclusion in plain English so that the reader can understand it even if that reader has never taken a statistics course.*

Let us practice these rules using the Chevy Impala Excel spreadsheet that you created earlier in this chapter, but first, we need to state the hypotheses for that car.

If General Motors wants to claim that the Chevy Impala gets 28 highway miles per gallon on a billboard ad, the hypotheses would be:

H_0: $\mu = 28$ mpg
H_1: $\mu \neq 28$ mpg

You will remember that the reference value of 28 mpg was inside the 95% confidence interval about the mean for your data, so we would accept H_0 for the Chevy Impala that the car does get 28 mpg.

Objective: To state the result when you accept H_0

Result: *Since the reference value of 28 mpg is inside the confidence interval, we accept the null hypothesis, H_0.*

Let us try our three rules now:

Objective: To write the conclusion when you accept H_0

Rule #1: *Since the reference value was inside the confidence interval, we cannot use the word "significantly" in the conclusion. This is a basic rule we are using in this chapter for every problem.*

Rule #2: The key terms in the conclusion would be:

- Chevy Impala
- reference value of 28 mpg

Rule #3: The Chevy Impala did get 28 mpg.

The process of writing the conclusion when you accept H_0 is relatively straightforward since you put into words what you said when you wrote the null hypothesis.

However, the process of stating the conclusion when you reject H_0 and accept H_1 is more difficult, so let us practice writing that type of conclusion with three practice case examples:

Objective: To write the result and conclusion when you reject H_0

Case #1: Suppose that an ad in *The Wall Street Journal* claimed that the Honda Accord Sedan gets 34 miles per gallon on the highway. The hypotheses would be:

H_0: $\mu = 34$ mpg
H_1: $\mu \neq 34$ mpg

Suppose that your research yields the following confidence interval:

	30		31		32		34	
	lower		Mean		upper		Ref.	
	limit				limit		Value	

Result: *Since the reference value is outside the confidence interval, we reject the null hypothesis and accept the research hypothesis.*

The three rules for stating the conclusion would be:

Rule #1: We must include the word "significantly" since the reference value of 34 is outside the confidence interval.

Rule #2: The key terms would be:

- Honda Accord Sedan
- significantly
- either "more than" or "less than"
- and probably closer to

Rule #3: The Honda Accord Sedan got significantly less than 34 mpg, and it was probably closer to 31 mpg.

Note that this conclusion says that the mpg was less than 34 mpg because the sample mean was only 31 mpg. Note, also, that when you find a significant result by rejecting the null hypothesis, *it is not sufficient to say only "significantly less than 34 mpg"* because that does not tell the reader "how much less than 34 mpg" the sample mean was from 34 mpg. To make the conclusion clear, you need to add "probably closer to 31 mpg" since the sample mean was only 31 mpg.

Case #2: Suppose that you have been hired as a consultant by the St. Louis Symphony Orchestra (SLSO) to analyze the data from an Internet survey of attendees for a concert in Powell Symphony Hall in St. Louis last month. You have decided to practice your data analysis skills on question #7 given in Fig. 3.8:

Question #7:	"Overall, how satisfied have you been with your experience(s) at SLSO concerts?"

1	2	3	4	5	6	7
Extremely dissatisfied						Extremely satisfied

Fig. 3.8 Example of a survey item used by the St. Louis Symphony Orchestra (SLSO)

The hypotheses for this one item would be:

H_0: $\mu = 4$
H_1: $\mu \neq 4$

Essentially, the null hypothesis equal to 4 states that if the obtained mean score for this question is not significantly different from 4 on the rating scale, then attendees, overall, were neither satisfied nor dissatisfied with their SLSO concerts.

Suppose that your analysis produced the following confidence interval for this item on the survey.

.... 1.8	2.8	3.8	4 ....
lower limit	Mean	upper limit	Ref. Value

Result: *Since the reference value is outside the confidence interval, we reject the null hypothesis and accept the research hypothesis.*

Rule #1: You must include the word "significantly" since the reference value is outside the confidence interval.

Rule #2: The key terms would be:

- attendees
- SLSO Internet survey
- significantly
- last month
- either satisfied or dissatisfied (since the result is significant)
- experiences at concerts
- overall

Rule #3: Attendees were significantly dissatisfied, overall, on last month's Internet survey with their experiences at concerts of the SLSO.

Note that you need to use the word "dissatisfied" since the sample mean of 2.8 was on the dissatisfied side of the middle of the rating scale.

Case #3: Suppose that Marriott Hotel at the St. Louis Airport location had the results of one item in its Guest Satisfaction Survey from last week's customers that was the following (see Fig. 3.9):

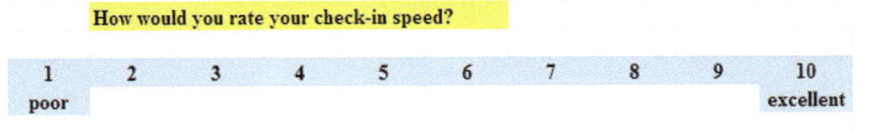

Fig. 3.9 Example of a Survey Item from Marriott Hotels

This item would have the following hypotheses:

H_0: $\mu = 5.5$
H_1: $\mu \neq 5.5$

Suppose that your research produced the following confidence interval for this item on the survey:

.... 5.5	5.7	5.8	5.9
Ref.	lower	Mean	upper
Value	limit		limit

Result: *Since the reference value is outside the confidence interval, we reject the null hypothesis and accept the research hypothesis.*

The three rules for stating the conclusion would be:

Rule #1: You must include the word "significantly" since the reference value is outside the confidence interval.

Rule #2: The key terms would be:

> – Marriott Hotel
> – St. Louis Airport
> – significantly
> – check-in speed
> – survey
> – last week
> – customers
> – either "positive" or "negative" (we will explain this)

Rule #3: Customers at the St. Louis Airport Marriott Hotel last week rated their check-in speed in a survey as significantly positive.

Note two important things about this conclusion above: (1) People when speaking English do not normally say "significantly excellent" since something is either excellent or is not excellent without any modifier, and (2) since the mean rating of the check-in speed (5.8) was significantly greater than 5.5 on the positive side of the scale, we would say "significantly positive" to indicate this fact.

The three practice problems at the end of this chapter will give you additional practice in stating the conclusion of your result, and this book will include many more examples that will help you to write a clear and accurate conclusion to your research findings.

3.3   Alternative Ways to Summarize the Result of a Hypothesis Test

It is important for you to understand that in this book we are summarizing a hypothesis test in one of two ways: (1) we accept the null hypothesis or (2) we reject the null hypothesis and accept the research hypothesis. We are consistent in the use of these words so that you can understand the concept underlying hypothesis testing.

However, there are many other ways to summarize the result of a hypothesis test, and all of them are correct theoretically, even though the terminology differs. If you are taking a course with a professor who wants you to summarize the results of a statistical test of hypotheses in language which is different from the language we are using in this book, do not panic! If you understand the concept of hypothesis testing as described in this book, you can then translate your understanding to use the terms that your professor wants you to use to reach the same conclusion to the hypothesis test.

Statisticians and professors of educational and psychological statistics all have their own language that they like to use to summarize the results of a hypothesis test. There is no one set of words that these statisticians and professors will ever agree on, and so we have chosen the one that we believe to be easier to understand in terms of the concept of hypothesis testing.

To convince you that there are many ways to summarize the results of a hypothesis test, we present the following quotes from prominent statistics and research books to give you an idea of the different ways that are possible.

3.3.1   Different Ways to Accept the Null Hypothesis

The following quotes are typical of the language used in statistics and research books when the null hypothesis is accepted:

> The null hypothesis is not rejected. (Black 2010, p. 310)
> The null hypothesis cannot be rejected. (McDaniel and Gates 2010, p. 545)
> The null hypothesis . . . claims that there is no difference between groups. (Salkind 2010, p. 193)
> The difference is not statistically significant. (McDaniel and Gates 2010, p. 545)
> . . . the obtained value is not extreme enough for us to say that the difference between Groups 1 and 2 occurred by anything other than chance. (Salkind 2010, p. 225)
> If we do not reject the null hypothesis, we conclude that there is not enough statistical evidence to infer that the alternative (hypothesis) is true. (Keller 2009, p. 358)
> The research hypothesis is not supported. (Zikmund and Babin 2010, p. 552)

3.3.2   Different Ways to Reject the Null Hypothesis

The following quotes are typical of the quotes used in statistics and research books when the null hypothesis is rejected:

> The null hypothesis is rejected. (McDaniel and Gates 2010, p. 546)
> If we reject the null hypothesis, we conclude that there is enough statistical evidence to infer that the alternative hypothesis is true. (Keller 2009, p. 358)
> If the test statistic's value is inconsistent with the null hypothesis, we reject the null hypothesis and infer that the alternative hypothesis is true. (Keller 2009, p. 348)
> Because the observed value . . . is greater than the critical value . . ., the decision is to reject the null hypothesis. (Black 2010, p. 359)
> If the obtained value is more extreme than the critical value, the null hypothesis cannot be accepted. (Salkind 2010, p. 243)
> The critical t-value . . . must be surpassed by the observed t-value if the hypothesis test is to be statistically significant . . . . (Zikmund and Babin 2010, p. 567)
> The calculated test statistic . . . exceeds the upper boundary and falls into this rejection region. The null hypothesis is rejected. (Weiers 2011, p. 330)

You should note that all of the above quotes are used by statisticians and professors when discussing the results of a hypothesis test, and so you should not be surprised if someone asks you to summarize the results of a statistical test using a different language than the one we are using in this book.

3.4 End-of-Chapter Practice Problems

1. Suppose that a school district wants to determine if the IQ scores of its current kindergarten students differ from the "rolling average" of its kindergarten students over the past 3 years. It decides to administer the Wechsler Preschool and Primary Scale for Intelligence-Revised (WPPSI-R) to its current kindergarten students at the same time of the school year that it administered this test over the past 3 years. The WPPSI-R attempts to measure the cognitive ability of children using 12 subtests, and the Full Scale IQ is scaled to have a mean of 100 and a standard deviation of 15. Over the past 3 years, the kindergarten students in this district have had a mean of 110. You have been asked to determine if the current year's kindergarten students differ in their IQ scores from the students over the past 3 years, and you decide to test your statistical skills by selecting a random sample of the current year's students. The hypothetical data appear in Fig. 3.10:

   (a) To the right of this table, use Excel to find the sample size, mean, standard deviation, and standard error of the mean for the IQ figures. Label your answers. Use number format (two decimal places) for the mean, standard deviation, and standard error of the mean.
   (b) Enter the null hypothesis and the research hypothesis onto your spreadsheet.
   (c) Use Excel's TINV function to find the 95% confidence interval about the mean for these figures. Label your answers. Use number format (two decimal places).
   (d) Enter your *result* onto your spreadsheet.
   (e) Enter your *conclusion in plain English* onto your spreadsheet.
   (f) Print the final spreadsheet to fit onto one page (if you need help remembering how to do this, see the objectives at the end of Chap. 2 in Sect. 2.4).
   (g) On your printout, draw a diagram of this 95% confidence interval by hand.
   (h) Save the file as: KINDER3.

2. Suppose that you have been asked by the Human Resources department (HR) at your company to analyze the data from a recent "morale survey" of its managers to find out how managers think about working at this company. You want to test out your Excel skills on a small sample of managers with one item from the survey. You select a random sample of managers, and the hypothetical data from item #24 are given in Fig. 3.11.
   Create an Excel spreadsheet with these data.

   (a) Use Excel to the right of the table to find the sample size, mean, standard deviation, and standard error of the mean for these data. Label your answers, and use two decimal places for the mean, standard deviation, and standard error of the mean.
   (b) Enter the null hypothesis and the research hypothesis for this item on your spreadsheet.

Sample of Kindergarten students for the WPPSI-R IQ test

IQ score
110
100
105
95
98
102
105
108
99
102
104
98
112
105
98
104
115
118
120
125
130
115

Fig. 3.10 Worksheet data for Chap. 3: practice problem #1

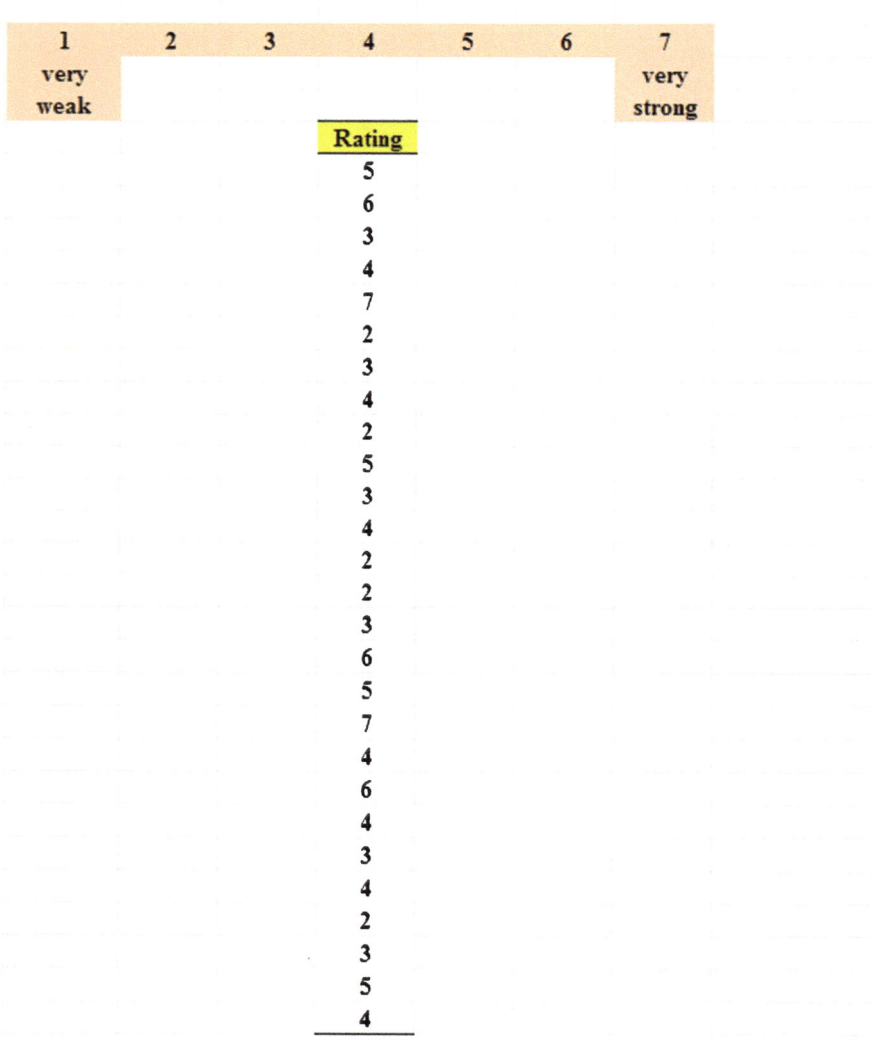

HUMAN RESOURCES DEPARTMENT

MORALE SURVEY OF MANAGERS

Item #24: **"How would you rate the quality of leadership shown by top management in this company?"**

1	2	3	4	5	6	7
very						very
weak						strong

Rating
5
6
3
4
7
2
3
4
2
5
3
4
2
2
3
6
5
7
4
6
4
3
4
2
3
5
4

Fig. 3.11 Worksheet data for Chap. 3: practice problem #2

(c) Use Excel's TINV function to find the 95% confidence interval about the mean for these data. Label your answers on your spreadsheet. Use two decimal places for the lower limit and the upper limit of the confidence interval.

(d) Enter the *result* of the test on your spreadsheet.

(e) Enter the *conclusion* of the test in plain English on your spreadsheet.

(f) Print your final spreadsheet so that it fits onto one page (if you need help remembering how to do this, see the objectives at the end of Chap. 2 in Sect. 2.4).

(g) Draw a picture of the confidence interval, including the reference value, onto your spreadsheet.

(h) Save the final spreadsheet as: top8.

3. Suppose that you have been asked to conduct three focus groups in different cities with adult women (ages 25–44) to determine how much they liked a new design of a blouse that was created by a well-known designer. The designer is hoping to sell this blouse in department stores at a retail price of $68.00. You conduct a 1-hour focus group discussion with three groups of adult women in this age range, and the last question on the survey at the end of the discussion period produced the hypothetical results given in Fig. 3.12:
Create an Excel spreadsheet with these data.

(a) Use Excel to the right of the table to find the sample size, mean, standard deviation, and standard error of the mean for these data. Label your answers, and use two decimal places and currency format for the mean, standard deviation, and standard error of the mean.

(b) Enter the null hypothesis and the research hypothesis for this item onto your spreadsheet.

(c) Use Excel's TINV function to find the 95% confidence interval about the mean for these data. Label your answers on your spreadsheet. Use two decimal places in currency format for the lower limit and the upper limit of the confidence interval.

(d) Enter the *result* of the test on your spreadsheet.

(e) Enter the *conclusion* of the test in plain English on your spreadsheet.

(f) Print your final spreadsheet so that it fits onto one page (if you need help remembering how to do this, see the objectives at the end of Chap. 2 in Sect. 2.4).

(g) Draw a picture of the confidence interval, including the reference value, onto your spreadsheet.

(h) Save the final spreadsheet as: blouse9.

Question #10:   "How much would you be willing to pay for this blouse?"

$ _____

Groups 1,2,3 in $
62
55
73
53
46
48
57
59
65
68
64
72
62
67
59
71
65
63
69
71
70
58
67
65
63
59
70
67
64
65

Fig. 3.12 Worksheet data for Chap. 3: practice problem #3

References

Black, K. Business Statistics: for Contemporary Decision Making (6th ed.). Hoboken, NJ: John Wiley& Sons, Inc., 2010.

Keller, G. Statistics for Management and Economics (8th ed.). Mason, OH: South-Western Cengage learning, 2009.

Levine, D.M. Statistics for Managers using Microsoft Excel (6th ed.). Boston, MA: Prentice Hall/ Pearson, 2011.

McDaniel, C. and Gates, R. Marketing Research (8th ed.). Hoboken, NJ: John Wiley & Sons, Inc., 2010.

Salkind, N.J. Statistics for People Who (think they) Hate Statistics (2nd Excel 2007 ed.). Los Angeles, CA: Sage Publications, 2010.

Weiers, R.M. Introduction to Business Statistics (7th ed.). Mason, OH: South-Western Cengage Learning, 2011.

Zikmund, W.G. and Babin, B.J. Exploring Marketing Research (10th ed.). Mason, OH: South-Western Cengage learning, 2010.

Chapter 4
One-Group *t*-Test for the Mean

In this chapter, you will learn how to use one of the most popular and most helpful statistical tests in educational and psychological research: the one-group *t*-test for the mean.

The formula for the one-group *t*-test is as follows:

$$t = \frac{\bar{X} - \mu}{S_{\bar{X}}} \quad \text{Where} \tag{4.1}$$

$$\text{s.e.} = S_{\bar{X}} = \frac{S}{\sqrt{n}} \tag{4.2}$$

This formula asks you to take the mean ($\bar{X}$) and subtract the population mean (μ) from it and then divide the answer by the standard error of the mean (s.e.). The standard error of the mean equals the standard deviation divided by the square root of *n* (the sample size).

Let us discuss the seven steps of hypothesis testing using the one-group *t*-test so that you can understand how this test is used.

4.1 The Seven Steps for Hypothesis Testing Using the One-Group *t*-Test

Objective: To learn the seven steps of hypothesis testing using the one-group *t*-test

T. Quirk, *Excel 2010 for Educational and Psychological Statistics:*
A Guide to Solving Practical Problems, DOI 10.1007/978-1-4614-2071-2_4,
© Springer Science+Business Media, LLC 2012

Before you can try out your Excel skills on the one-group *t*-test, you need to learn the basic steps of hypothesis testing for this statistical test. There are seven steps in this process:

4.1.1 Step 1: State the Null Hypothesis and the Research Hypothesis

If you are using numerical scales in your survey, you need to remember that these hypotheses refer to the "middle" of the numerical scale. For example, if you are using seven-point scales with $1 =$ poor and $7 =$ excellent, these hypotheses would refer to the middle of these scales and would be:

Null hypothesis H_0: $\mu = 4$
Research hypothesis H_1: $\mu \neq 4$

As a second example, suppose that you worked for Honda Motor Company and that you wanted to place a magazine ad that claimed that the new Honda Fit got 35 miles per gallon (mpg). The hypotheses for testing this claim on actual data would be:

H_0: $\mu = 35$ mpg
H_1: $\mu \neq 35$ mpg

4.1.2 Step 2: Select the Appropriate Statistical Test

In this chapter, we will be studying the one-group *t*-test, and so we will select that test.

4.1.3 Step 3: Decide on a Decision Rule for the One-Group t-Test

(a) If the absolute value of *t* is less than the critical value of *t*, accept the null hypothesis.
(b) If the absolute value of *t* is greater than the critical value of *t*, reject the null hypothesis and accept the research hypothesis.

You are probably saying to yourself: "That sounds fine, but how do I find the absolute value of *t*?"

4.1.3.1 Finding the Absolute Value of a Number

To do that, we need another objective.

Objective: To find the absolute value of a number

If you took a basic algebra course in high school, you may remember the concept of "absolute value." In mathematical terms, the absolute value of any number is *always* that number expressed as a positive number.

For example, the absolute value of 2.35 is +2.35.

And the absolute value of minus 2.35 (i.e., -2.35) is also +2.35.

This becomes important when you are using the *t*-table in Appendix E of this book. We will discuss this table later when we get to step 5 of the one-group *t*-test where we explain how to find the critical value of *t* using Appendix E.

4.1.4   Step 4: Calculate the Formula for the One-Group t-Test

---

Objective: To learn how to use the formula for the one-group *t*-test

---

The formula for the one-group *t*-test is as follows:

$$t = \frac{\bar{X} - \mu}{S_{\bar{X}}} \quad \text{Where} \qquad (4.1)$$

$$\text{s.e.} = S_{\bar{X}} = \frac{S}{\sqrt{n}} \qquad (4.2)$$

This formula makes the following assumptions about the data (Foster et al. 1998): (1) The data are independent of each other (i.e., each person receives only one score), (2) the *population* of the data is normally distributed, and (3) the data have a constant variance (note that the standard deviation is the square root of the variance).

To use this formula, you need to follow these steps:

1. Take the sample mean in your research study and subtract the population mean μ from it (remember that the population mean for a study involving numerical rating scales is the "middle" number in the scale).
2. Then take your answer from the above step and divide your answer by the standard error of the mean for your research study (you will remember that you learned how to find the standard error of the mean in Chap. 1; to find the standard error of the mean, just take the standard deviation of your research study and divide it by the square root of *n*, where *n* is the number of people used in your research study).
3. The number you get after you complete the above step is the value for *t* that results when you use the formula stated above.

4.1.5   Step 5: Find the Critical Value of t in the t-Table in Appendix E

Objective: To find the critical value of *t* in the *t*-table in Appendix E

Before we get into an explanation of what is meant by "the critical value of *t*," let us give you practice in finding the critical value of *t* by using the *t*-table in Appendix E.

Keep your finger on Appendix E as we explain how you need to "read" that table.

Since the test in this chapter is called the "one-group *t*-test," you will use the first column on the left in Appendix E to find the critical value of *t* for your research study (note that this column is headed: "*n*").

To find the critical value of *t*, you go down to this first column until you find the sample size in your research study, and then you go to the right and read the critical value of *t* for that sample size in the critical *t* column in the table (note that *this is the column that you would use for both the one-group t-test and the 95% confidence interval about the mean*).

For example, if you have 27 people in your research study, the critical value of *t* is 2.056.

If you have 38 people in your research study, the critical value of *t* is 2.026.

If you have more than 40 people in your research study, the critical value of *t* is always 1.96.

Note that the "critical *t* column" in Appendix E represents the value of *t* that you need to obtain to be 95% confident of your results as "significant" results.

The critical value of *t* is the value that tells you whether or not you have found a "significant result" in your statistical test.

The *t*-table in Appendix E represents a series of "bell-shaped normal curves" (they are called bell-shaped because they look like the outline of the Liberty Bell that you can see in Philadelphia outside of Independence Hall).

The "middle" of these normal curves is treated as if it were zero point on the x-axis (the technical explanation of this fact is beyond the scope of this book, but any good statistics book (e.g., Zikmund and Babin 2010) will explain this concept to you if you are interested in learning more about it).

Thus, values of *t* that are to the right of this zero point are positive values that use a plus sign before them, and values of *t* that are to the left of this zero point are negative values that use a minus sign before them. Thus, some values of *t* are positive, and some are negative.

However, every statistics book that includes a *t*-table only reprints the *positive* side of the *t*-curves because the negative side is the mirror image of the positive side; this means that the negative side contains the exact same numbers as the positive side, but the negative numbers all have a minus sign in front of them.

Therefore, to use the *t*-table in Appendix E, you need to *take the absolute value of the t-value you found when you use the t-test formula* since the *t*-table in Appendix E only has the values of *t* that are the positive values for *t*.

Throughout this book, we are assuming that you want to be 95% confident in the results of your statistical tests. Therefore, the value for *t* in the *t*-table in Appendix E tells you whether or not the *t*-value you obtained when you used the formula for the one-group *t*-test is within the 95% interval of the *t*-curve range that that *t*-value would be expected to occur with 95% confidence.

If the *t*-value you obtained when you used the formula for the one-group *t*-test is *inside* of the 95% confidence range, we say that the result you found is *not significant* (note that this is equivalent to *accepting the null hypothesis!*).

If the *t*-value you found when you used the formula for the one-group *t*-test is *outside* of this 95% confidence range, we say that you have found a *significant result* that would be expected to occur less than 5% of the time (note that this is equivalent to *rejecting the null hypothesis and accepting the research hypothesis*).

4.1.6   Step 6: State the Result of Your Statistical Test

There are two possible results when you use the one-group *t*-test, and only one of them can be accepted as "true."

Either: Since the absolute value of *t* that you found in the *t*-test formula is *less than the critical value of t* in Appendix E, you accept the null hypothesis.

Or: Since the absolute value of *t* that you found in the *t*-test formula is *greater than the critical value of t* in Appendix E, you reject the null hypothesis and accept the research hypothesis.

4.1.7   Step 7: State the Conclusion of Your Statistical Test in Plain English!

In practice, this is more difficult than it sounds because you are trying to summarize the result of your statistical test in simple English that is both concise and accurate so that someone who has never had a statistics course (such as your boss, perhaps) can understand the result of your test. This is a difficult task, and we will give you lots of practice doing this last and most important step throughout this book.

If you have read this far, you are ready to sit down at your computer and perform the one-group *t*-test using Excel on some hypothetical data from the Guest Satisfaction Survey used by Marriott Hotels.

Let us give this a try.

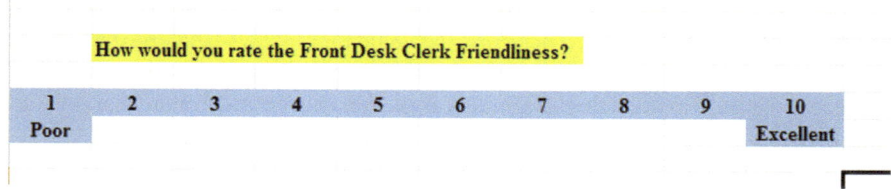

Fig. 4.1 Sample survey item for Marriot Hotel (practical example)

4.2   One-Group *t*-Test for the Mean

Suppose that you have been hired as a statistical consultant by Marriott Hotel in St. Louis to analyze the data from a Guest Satisfaction survey that they give to all customers to determine the degree of satisfaction of these customers for various activities of the hotel.

The survey contains a number of items, but suppose item #7 is the one in Fig. 4.1.

Suppose, further, that you have decided to analyze the data from last week's customers using the one-group *t*-test.

Important note: *You would need to use this test for each of the survey items separately.*

Suppose that the hypothetical data for item #7 from last week at the St. Louis Marriott Hotel were based on a sample size of 124 guests who had a mean score on this item of 6.58 and a standard deviation on this item of 2.44.

> Objective: To analyze the data for each question separately using the one-group *t*-test for each survey item

Create an Excel spreadsheet with the following information:

B11:   Null hypothesis
B14:   Research hypothesis

Note: *Remember that when you are using a rating scale item, both the null hypothesis and the research hypothesis refer to the "middle of the scale." In the ten-point scale in this example, the middle of the scale is 5.5 since five numbers are below 5.5 (i.e., 1–5) and five numbers are above 5.5 (i.e., 6–10). Therefore, the hypotheses for this rating scale item are:*

H_0: $\mu = 5.5$
H_1: $\mu \neq 5.5$
B17:   *n*
B20:   mean
B23:   STDEV
B26:   s.e.
B29:   critical *t*

B32:   *t*-Test
B36:   Result:
B41:   Conclusion:

   Now, use Excel:

D17:   Enter the sample size.
D20:   Enter the mean.
D23:   Enter the STDEV (see Fig. 4.2).

Null hypothesis:	
Research hypothesis:	
n	124
mean	6.58
STDEV	2.44
s.e.	
critical t	
t-test	
Result:	
Conclusion:	

Fig. 4.2 Basic data table for
front desk clerk friendliness

D26   Compute the standard error using the formula in Chap. 1.
D29   Find the critical *t* value of *t* in the *t*-table in Appendix E.

Now, enter the following formula in cell D32 to find the *t*-test result:

=(D20–5.5)/D26.

This formula takes the sample mean (D20) and subtracts the population hypothesized mean of 5.5 from the sample mean and *then* divides the answer by the standard error of the mean (D26). Note that you need to enter D20–5.5 with an open parenthesis *before* D20 and a closed parenthesis *after* 5.5 so that the *answer of 1.08 is then divided by the standard error of 0.22* to get the *t*-test result of 4.93.

Now, use two decimal places for both the s.e. and the *t*-test result (see Fig. 4.3).

Null hypothesis:	
Research hypothesis:	
n	124
mean	6.58
STDEV	2.44
s.e.	0.22
critical t	1.96
t-test	4.93
Result:	
Conclusion:	

Fig. 4.3 *t*-Test formula result for front desk clerk friendliness

Now, write the following sentence in D36:D39 to summarize the result of the *t*-test:

D36:   Since the absolute value of *t* of 4.93 is
D37:   greater than the critical *t* of 1.96, we
D38:   reject the null hypothesis and accept
D39:   the research hypothesis.

Lastly, write the following sentence in D41:D43 to summarize the conclusion of the result for item #7 of the Marriott Guest Satisfaction Survey:

D41:   St. Louis Marriott Hotel guests rated the
D42:   front desk clerks as significantly
D43:   friendly last week.

Save your file as: MARRIOTT3

Important note: *You are probably wondering why we entered both the result and the conclusion in separate cells instead of in just one cell. This is because if you enter them in one cell, you will be very disappointed when you print out your final spreadsheet, because one of two things will happen that you will not like: (1) if you print the spreadsheet to fit onto only one page, the result and the conclusion will force the entire spreadsheet to be printed in such small font size that you will be unable to read it, or (2) if you do not print the final spreadsheet to fit onto one page, both the result and the conclusion will "dribble over" onto a second page instead of fitting the entire spreadsheet onto one page. In either case, your spreadsheet will not have a "professional look."*

Print the final spreadsheet so that it fits onto one page as given in Fig. 4.4. Enter the null hypothesis and the research hypothesis by hand on your spreadsheet.

Important note: *It is important for you to understand that "technically" the above conclusion in statistical terms should state:*
   "St. Louis Marriott Hotel Guests rated the Front Desk Clerks as friendly last week, and this result was probably not obtained by chance."

   However, throughout this book, we are using the term "significantly" in writing the conclusion of statistical tests to alert the reader that the result of the statistical test was probably not a chance finding, but instead of writing all of those words each time, we use the word "significantly" as a shorthand to the longer explanation. This makes it much easier for the reader to understand the conclusion when it is written "in plain English," instead of technical, statistical language.

Null hypothesis:	$\mu = 5.5$
Research hypothesis:	$\mu \neq 5.5$
n	124
mean	6.58
STDEV	2.44
s.e.	0.22
critical t	1.96
t-test	4.93
Result:	Since the absolute value of t of 4.93 is greater than the critical t of 1.96, we reject the null hypothesis and accept the research hypothesis.
Conclusion:	St. Louis Marriott Hotel guests rated the Front Desk Clerks as significantly friendly last week.

Fig. 4.4 Final spreadsheet for front desk clerk friendliness

4.3   Can You Use Either the 95% Confidence Interval About the Mean or the One-Group *t*-Test When Testing Hypotheses?

You are probably asking yourself:

"It sounds like you could use *either* the 95% confidence interval about the mean *or* the one-group *t*-test to analyze the results of the types of problems described so far in this book? Is this a correct statement?"

The answer is a resounding: *"Yes!"*

Both the confidence interval about the mean and the one-group *t*-test are used often in educational and psychological research on the types of problems described so far in this book. *Both of these tests produce the same result and the same conclusion from the data set!*

Both of these tests are explained in this book because some researchers prefer the confidence interval about the mean test, others prefer the one-group *t*-test, and still others prefer to use both tests on the same data to make their results and conclusions clearer to the reader of their research reports. Since we do not know which of these tests your researcher prefers, we have explained both of them so that you are competent in the use of both tests in the analysis of statistical data.

Now, let us try your Excel skills on the one-group *t*-test on these three problems at the end of this chapter.

4.4   End-of-Chapter Practice Problems

1. Subaru of America rates the customer satisfaction of its dealers on a weekly basis on its Purchase Experience Survey and demands that dealers achieve a 93% satisfaction score, or the dealers are required to take additional training to improve their customer satisfaction scores. Suppose that you have selected a random sample of rating forms submitted by new car purchasers (either online or through the mail) for the St. Louis Subaru dealer in the past week and that you have prepared the hypothetical table in Fig. 4.5 for question #1d:

   (a) Write the null hypothesis and the research hypothesis on your spreadsheet.
   (b) Use Excel to find the sample size, mean, standard deviation, and standard error of the mean to the right of the data set. Use number format (two decimal places) for the mean, standard deviation, and standard error of the mean.
   (c) Enter the critical *t* from the *t*-table in Appendix E onto your spreadsheet and label it.
   (d) Use Excel to compute the *t*-value for these data (use 2 decimal places) and label it on your spreadsheet.
   (e) Type the result on your spreadsheet and then type the conclusion in plain English on your spreadsheet.
   (f) Save the file as: subaru4.

2. Suppose that the mathematics department chair at a university wants to know if this year's students enrolled in a first semester calculus course are similar in mathematical knowledge to students from the previous year. This university gives a standardized math test to all students who take calculus at the beginning of the course to assess their math skills. Last year's students had an average score of 88 on this exam. The hypothetical data for a random sample of students from this year's exam is presented in Fig. 4.6:

   (a) *On your Excel spreadsheet*, write the null hypothesis and the research hypothesis for these data.

SUBARU Customer Satisfaction Survey
Results for the week of October 2, 2011
St. Louis Subaru dealer on Big Bend

Item #1d: "The salesperson was knowledgeable about the Subaru model line."

Score	Rating
1	Completely Disagree
2	Disagree
3	Somewhat Disagree
4	Neither Agree nor Disagree
5	Somewhat Agree
6	Agree
7	Completely Agree

Rating
5
7
6
4
3
5
6
7
2
3
5
7
4
7
7
5
6
6
4
3
5
5

Fig. 4.5 Worksheet data for Chap. 4: practice problem #1

(b) Use Excel to find the *sample size, mean, standard deviation, and standard error of the mean* for these data (two decimal places for the mean, standard deviation, and standard error of the mean).
(c) Use Excel to perform a *one-group t-test* on these data (two decimal places).
(d) On your printout, type the *critical value of t* (.05 level) given in your *t*-table in Appendix E.
(e) On your spreadsheet, type the *result* of the *t*-test.
(f) On your spreadsheet, type the *conclusion* of your study in plain English.
(g) Save the file as: calculus4

First-semester Calculus
Standardized math test

Score
87
90
85
94
93
88
82
85
96
89
92
94
91
89
90
92
93
89
88
87
86
92
94
95
88

Fig. 4.6 Worksheet data for Chap. 4: practice problem #2

3. Suppose that you have been hired as a marketing consultant by the Missouri Botanical Garden and have been asked to redesign the Comment Card survey that they have been asking visitors to the Garden to fill out after their visit. The Garden has been using a five-point rating scale with 1 = poor and 5 = excellent. Suppose, further, that you have convinced the Garden staff to change to a nine-point scale with 1 = poor and 9 = excellent so that the data will have a

Missouri Botanical Garden

Results of the survey of Nov. 6, 2011

Item #10:   "How would you rate the helpfulness of The Garden staff?"

1	2	3	4	5	6	7	8	9
Poor								Excellent

Rating
8
6
5
7
9
5
6
4
8
7
6
8
6
7
9
7
6
3
8
7
6

Fig. 4.7 Worksheet data for Chap. 4: practice problem #3

larger standard deviation. The hypothetical results of a recent week for question #10 of your revised survey appear in Fig. 4.7:

(a) Write the null hypothesis and the research hypothesis on your spreadsheet.
(b) Use Excel to find the sample size, mean, standard deviation, and standard error of the mean to the right of the data set. Use number format (2 decimal places) for the mean, standard deviation, and standard error of the mean.
(c) Enter the critical *t* from the *t*-table in Appendix E onto your spreadsheet and label it.
(d) Use Excel to compute the *t*-value for these data (use two decimal places) and label it on your spreadsheet.
(e) Type the result on your spreadsheet and then type the conclusion in plain English on your spreadsheet
(f) Save the file as: Garden5.

References

Zikmund, W.G. and Babin, B.J. Exploring Marketing Research (10th ed.) Mason, OH: South-Western Cengage Learning, 2010.

Foster, D.P., Stine, R.A., Waterman, R.P. Basic Business Statistics: A Casebook. New York, NY: Springer-Verlag, 1998.

Chapter 5
Two-Group *t*-Test of the Difference of the Means for Independent Groups

Up until now in this book, you have been dealing with the situation in which you have had only one group of people in your research study and only one measurement "number" on each of these people. We will now change gears and deal with the situation in which you are measuring two groups of people instead of only one group of people.

Whenever you have two completely different groups of people (i.e., no one person is in both groups, but every person is measured on only one variable to produce one "number" for each person), we say that the two groups are "independent of one another." This chapter deals with just that situation, and that is why it is called the two-group *t*-test for independent groups.

The two assumptions underlying the two-group *t*-test are the following (Zikmund and Babin 2010): (1) both groups are sampled from a normal population, and (2) the variances of the two populations are approximately equal. Note that the standard deviation is merely the square root of the variance. (There are different formulas to use when each person is measured twice to create two groups of data, and this situation is called "dependent," but those formulas are beyond the scope of this book.) This book only deals with two groups that are independent of one another so that no person is in both groups of data.

When you are testing for the difference between the means for two groups, it is important to remember that there are two different formulas that you need to use depending on the sample sizes of the two groups:

1. Use Formula #1 in this chapter when both of the groups have more than 30 people in them.
2. Use Formula #2 in this chapter when either one group or both groups have sample sizes less than 30 people in them.

We will illustrate both of these situations in this chapter.

But first, we need to understand the steps involved in hypothesis testing when two groups of people are involved before we dive into the formulas for this test.

T. Quirk, *Excel 2010 for Educational and Psychological Statistics:*
A Guide to Solving Practical Problems, DOI 10.1007/978-1-4614-2071-2_5,
© Springer Science+Business Media, LLC 2012

5.1   The Nine Steps for Hypothesis Testing Using the Two-Group *t*-Test

> Objective: To learn the nine steps of hypothesis testing using two groups of people and the two-group *t*-test

You will see that these steps parallel the steps used in the previous chapter that dealt with the one-group *t*-test, but there are some important differences between the steps that you need to understand clearly before we dive into the formulas for the two-group *t*-test.

5.1.1   Step 1: Name One Group, Group 1, and the Other Group, Group 2

The formulas used in this chapter will use the numbers 1 and 2 to distinguish between the two groups. If you define which group is group 1 and which group is group 2, you can use these numbers in your computations without having to write out the names of the groups.

For example, if you are testing teenage boys on their preference for the taste of Coke or Pepsi, you could call the groups "Coke" and "Pepsi," but this would require your writing out the words "Coke" or "Pepsi" whenever you wanted to refer to one of these groups. If you call the Coke group, Group 1, and the Pepsi group, Group 2, this makes it much easier to refer to the groups because it saves your writing time.

As a second example, you could be comparing the test market results for Kansas City vs. Indianapolis, but if you had to write out the names of those cities whenever you wanted to refer to them, it would take you more time than it would if, instead, you named one city, group 1, and the other city, Group 2.

Note, also, that it is completely arbitrary which group you call Group 1 and which group you call Group 2. You will achieve the same result and the same conclusion from the formulas however you decide to define these two groups.

5.1.2   Step 2: Create a Table That Summarizes the Sample Size, Mean Score, and Standard Deviation of Each Group

This step makes it easier for you to make sure that you are using the correct numbers in the formulas for the two-group *t*-test. If you get the numbers "mixed-up," your entire formula work will be incorrect, and you will botch the problem terribly.

For example, suppose that you tested teenage boys on their preference for the taste of Coke vs. Pepsi in which the boys were randomly assigned to taste just one of these brands and then rate its taste on a 100-point scale from 0 = poor to 100 = excellent. After the research study was completed, suppose that the Coke group had 52 boys in it, their mean taste rating was 55 with a standard deviation of 7, while the Pepsi group had 57 boys in it, and their average taste rating was 64 with a standard deviation of 13.

The formulas for analyzing these data to determine if there was a significant difference in the taste rating for teenage boys for these two brands require you to use six numbers correctly in the formulas: the sample size, the mean, and the standard deviation of each of the two groups. All six of these numbers must be used correctly in the formulas if you are to analyze the data correctly.

If you create a table to summarize these data, a good example of the table, using both Step 1 and Step 2, would be the data presented in Fig. 5.1.

Group	n	Mean	STDEV
1 (name it)			
2 (name it)			

Fig. 5.1 Basic table format for the two-group *t*-test

For example, if you decide to call Group 1 the Coke group and Group 2 the Pepsi group, the following table would place the six numbers from your research study into the proper calls of the table as in Fig. 5.2

Group	n	Mean	STDEV
1 (name it)	52	55	7
2 (name it)	57	64	13

Fig. 5.2 Results of entering the data needed for the two-group *t*-test

You can now use the formulas for the two-group *t*-test with more confidence that the six numbers will be placed in the proper place in the formulas.

Note that you could just as easily call Group 1 the Pepsi group and Group 2 the Coke group; it makes no difference how you decide to name the two groups; this decision is up to you

5.1.3 Step 3: State the Null Hypothesis and the Research Hypothesis for the Two-Group t-Test

If you have completed Step 1 above, this step is very easy because the null hypothesis and the research hypothesis will always be stated in the same way for the two-group *t*-test. The null hypothesis states that the population means of the two groups are equal, while the research hypothesis states that the population means of the two groups are not equal. In notation format, this becomes:

H_0: $\mu_1 = \mu_2$
H_1: $\mu_1 \neq \mu_2$

You can now see that this notation is much simpler than having to write out the names of the two groups in all of your formulas.

5.1.4 Step 4: Select the Appropriate Statistical Test

Since this chapter deals with the situation in which you have two groups of people but only one measurement on each person in each group, we will use the two-group *t*-test throughout this chapter.

5.1.5 Step 5: Decide on a Decision Rule for the Two-Group t-Test

The decision rule is exactly what it was in the previous chapter (see Sect. 4.1.3) when we dealt with the one-group *t*-test.

(a) If the absolute value of *t* is less than the critical value of *t*, accept the null hypothesis.
(b) If the absolute value of *t* is greater than the critical value of *t*, reject the null hypothesis and accept the research hypothesis.

Since you learned how to find the absolute value of *t* in the previous chapter (see Sect. 4.1.3.1), you can use that knowledge in this chapter.

5.1.6 Step 6: Calculate the Formula for the Two-Group t-Test

Since we are using two different formulas in this chapter for the two-group *t*-test depending on the sample size of the people in the two groups, we will explain how to use those formulas later in this chapter.

5.1.7   Step 7: Find the Critical Value of t in the t-Table in Appendix E

In the previous chapter where we were dealing with the one-group t-test, you found the critical value of t in the t-table in Appendix E by finding the sample size for the one group of people in the first column of the table and then reading the critical value of t across from it on the right in the "critical t column" in the table (see Sect. 4.1.5). This process was fairly simple once you have had some practice in doing this step.

However, for the two-group t-test, the procedure for finding the critical value of t is more complicated because you have two different groups of people in your study, and they often have different sample sizes in each group.

To use Appendix E correctly in this chapter, you need to learn how to find the "degrees of freedom" for your study. We will discuss that process now.

5.1.7.1   Finding the Degrees of Freedom (df) for the Two-Group t-Test

Objective: To find the degrees of freedom for the two-group t-test and to use it to find the critical value of t in the t-table in Appendix E

The mathematical explanation of the concept of the "degrees of freedom" is beyond the scope of this book, but you can find out more about this concept by reading any good statistics book (e.g., Keller 2009). For our purposes, you can easily understand how to find the degrees of freedom and to use it to find the critical value of t in Appendix E. The formula for the degrees of freedom (df) is:

$$\text{degrees of freedom} = \text{df} = n_1 + n_2 - 2 \qquad (5.1)$$

In other words, you add the sample size for Group 1 to the sample size for Group 2 and then subtract 2 from this total to get the number of degrees of freedom to use in Appendix E.

Take a look at Appendix E.

Instead of using the first column as we did in the one-group t-test that is based on the sample size, n, of one group of people, we need to use the second column of this table (df) to find the critical value of t for the two-group t-test.

For example, if you had 13 people in Group 1 and 17 people in Group 2, the degrees of freedom would be $13 + 17 - 2 = 28$, and the critical value of t would be 2.048 *since you look down the second column which contains the degrees of freedom* until you come to the number 28, and then read 2.048 in the "critical t column" in the table to find the critical value of t when df $= 28$.

As a second example, if you had 52 people in Group 1 and 57 people in Group 2, the degrees of freedom would be $52 + 57 - 2 = 107$. When you go down the second column in Appendix E for the degrees of freedom, you find that *once you go beyond the degrees of freedom equal to 39, the critical value of t is always 1.96* and that is the value you would use for the critical *t* with this example.

5.1.8 Step 8: State the Result of Your Statistical Test

The result follows the exact same result format that you found for the one-group *t*-test in the previous chapter (see Sect. 4.1.6).

Either: Since the absolute value of *t* that you found in the *t*-test formula is *less than the critical value of t* in Appendix E, you accept the null hypothesis.

Or: Since the absolute value of *t* that you found in the *t*-test formula is *greater than the critical value of t* in Appendix E, you reject the null hypothesis and accept the research hypothesis.

5.1.9 Step 9: State the Conclusion of Your Statistical Test in Plain English!

Writing the conclusion for the two-group *t*-test is more difficult than writing the conclusion for the one-group *t*-test because you have to decide what the difference was between the two groups.

When you accept the null hypothesis, the conclusion is simple to write: "There is no difference between the two groups in the variable that was measured."

But when you reject the null hypothesis and accept the research hypothesis, you need to be careful about writing the conclusion so that it is both accurate and concise.

Let us give you some practice in writing the conclusion of a two-group *t*-test.

5.1.9.1 Writing the Conclusion of the Two-Group *t*-Test When You Accept the Null Hypothesis

Objective: To write the conclusion of the two-group *t*-test when you have accepted the null hypothesis

Suppose that you have been hired as a statistical consultant by Marriott Hotel in St. Louis to analyze the data from a Guest Satisfaction Survey that they give to all customers to determine the degree of satisfaction of these customers for various activities of the hotel.

The survey contains a number of items, but suppose item #7 is the one in Fig. 5.3.

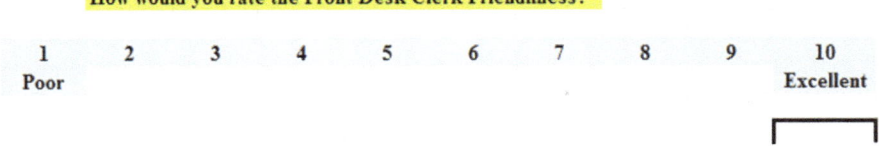

How would you rate the Front Desk Clerk Friendliness?

1	2	3	4	5	6	7	8	9	10
Poor									Excellent

Fig. 5.3 Marriott Hotel guest satisfaction survey item #7

Suppose further that you have decided to analyze the data from last week's customers comparing men and women using the two-group *t*-test.

Important note: *You would need to use this test for each of the survey items separately.*

Suppose that the hypothetical data for item #7 from last week at the St. Louis Marriott Hotel were based on a sample size of 124 men who had a mean score on this item of 6.58 and a standard deviation on this item of 2.44. Suppose that you also had data from 86 women from last week who had a mean score of 6.45 with a standard deviation of 1.86.

We will explain later in this chapter how to produce the results of the two-group *t*-test using its formulas, but, for now, let us "cut to the chase" and tell you that those formulas would produce the following in Fig. 5.4.

⊿	A	B	C	D	E	F
1						
2						
3	Group		n	Mean	STDEV	
4	1 Males		124	6.58	2.44	
5	2 Females		86	6.45	1.86	
6						
7						

Fig. 5.4 Worksheet data for males vs. females for the St. Louis Marriott Hotel for accepting the null hypothesis

degrees of freedom:	208
critical *t*:	1.96 (in Appendix E)
t-test formula:	0.44 (when you use your calculator!)
Result:	Since the absolute value of 0.44 is less than the critical *t* of 1.96, we accept the null hypothesis.
Conclusion:	There was no difference between male and female guests last week in their rating of the friendliness of the front desk clerk at the St. Louis Marriott Hotel.

Now, let us see what happens when you reject the null hypothesis (H_0) and accept the research hypothesis (H_1).

5.1.9.2   Writing the Conclusion of the Two-Group *t*-Test When You Reject the Null Hypothesis and Accept the Research Hypothesis

> Objective: To write the conclusion of the two-group *t*-test when you have rejected the null hypothesis and accepted the research hypothesis

Let us continue with this same example of the Marriott Hotel, but with the result that we reject the null hypothesis and accept the research hypothesis.

Let us assume that this time, you have data on 85 males from last week, and their mean score on this question was 7.26 with a standard deviation of 2.35. Let us further suppose that you also have data on 48 females from last week, and their mean score on this question was 4.37 with a standard deviation of 3.26.

Without going into the details of the formulas for the two-group *t*-test, these data would produce the following result and conclusion based on Fig. 5.5.

	A	B		C	D	E	
1							
2							
3		Group		n	Mean	STDEV	
4		1 Males		85	7.26	2.35	
5		2 Females		48	4.37	3.26	
6							
7							

Fig. 5.5 Worksheet data for St. Louis Marriott Hotel for obtaining a significant difference between males and females

Null Hypothesis:	$\mu_1 = \mu_2$
Research Hypothesis:	$\mu_1 \neq \mu_2$
degrees of freedom:	131
critical *t*:	1.96 (in Appendix E)
t-test formula:	5.40 (when you use your calculator!)
Result:	Since the absolute value of 5.40 is greater than the critical *t* of 1.96, we reject the null hypothesis and accept the research hypothesis.

Now, you need to compare the ratings of the men and women to find out which group had the more positive rating of the friendliness of the front desk clerk using the following rule:

Rule: *To summarize the conclusion of the two-group t-test, just compare the means of the two groups, and be sure to use the word "significantly" in your conclusion if you rejected the null hypothesis and accepted the research hypothesis.*

A good way to prepare to write the conclusion of the two-group *t*-test when you are using a rating scale is to place the mean scores of the two groups on a drawing of the scale so that you can visualize the difference of the mean scores. For example, for our Marriott Hotel example above, you would draw this "picture" of the scale in Fig. 5.6.

Fig. 5.6   Example of drawing a "picture" of the means of the two groups on the rating scale

This drawing tells you visually that males had a higher positive rating than females on this item (7.26 vs. 4.37). *And since you rejected the null hypothesis and accepted the research hypothesis, you know that you have found a significant difference between the two mean scores.*

So, our conclusion needs to contain the following key words:

- male guests
- female guests
- Marriott Hotel
- St. Louis
- last week
- significantly
- front desk clerks
- more friendly *or* less friendly
- *either* (7.26 vs. 4.37) *or* (4.37 vs. 7.26)

We can use these key words to write the either of two conclusions which are *logically identical*.

Either: Male guests at the Marriott Hotel in St. Louis last week rated the front desk clerks as significantly more friendly than female guests (7.26 vs. 4.37).

Or: Female guests at the Marriott Hotel in St. Louis last week rated the front desk clerks as significantly less friendly than male guests (4.37 vs. 7.26).

Both of these conclusions are accurate, so you can decide which one you want to write. It is your choice.

Also, note that the mean scores in parentheses at the end of these conclusions must match the sequence of the two groups in your conclusion. For example, if you say that: "Male guests rated the front desk clerks as significantly more friendly than female guests," the end of this conclusion should be (7.26 vs. 4.37) since you mentioned males first and females second.

Alternately, if you wrote that: "Female guests rated the front desk clerks as significantly less friendly than male guests," the end of this conclusion should be (4.37 vs. 7.26) since you mentioned females first and males second.

Putting the two mean scores at the end of your conclusion saves the reader from having to turn back to the table in your research report to find these mean scores to see how far apart the mean scores were.

Now, let us discuss Formula #1 that deals with the situation in which both groups have more than 30 people in them.

> Objective: To use Formula #1 for the two-group *t*-test when both groups have a sample size greater than 30 people

5.2 Formula #1: Both Groups Have More Than 30 People in Them

The first formula we will discuss will be used when you have two groups of people with more than 30 people in each group and one measurement on each person in each group. This formula for the two-group *t*-test is:

$$t = \frac{\bar{X}_1 - \bar{X}_2}{S_{\bar{X}_1 - \bar{X}_2}} \qquad (5.2)$$

$$\text{where } S_{\bar{X}_1 - \bar{X}_2} = \sqrt{\frac{S_1^2}{n_1} + \frac{S_2^2}{n_2}} \qquad (5.3)$$

$$\text{and where degrees of freedom} = \text{df} = n_1 + n_2 - 2 \qquad (5.1)$$

This formula looks daunting when you first see it, but let us explain some of the parts of this formula.

We have explained the concept of "degrees of freedom" earlier in this chapter, and so you should be able to find the degrees of freedom needed for this formula in order to find the critical value of *t* in Appendix E.

In the previous chapter, *the formula for the one-group t-test was the following*:

$$t = \frac{\bar{X} - \mu}{S_{\bar{X}}} \qquad (4.1)$$

$$\text{where s.e.} = S_{\bar{X}} = \frac{S}{\sqrt{n}} \qquad (4.2)$$

For the one-group t-test, you found the mean score and subtracted the population mean from it and then divided the result by the standard error of the mean (s.e.) to get the result of the t-test. You then compared the t-test result to the critical value of t to see if you either accepted the null hypothesis or rejected the null hypothesis and accepted the research hypothesis.

The two-group t-test requires a different formula because you have two groups of people, each with a mean score on some variable. You are trying to determine whether to accept the null hypothesis that the *population means of the two groups are equal* (in other words, there is no difference statistically between these two means), or whether the difference between the means of the two groups is "sufficiently large" that you would accept *that there is a significant difference* in the mean scores of the two groups.

The numerator of the two-group t-test asks you to find the difference of the means of the two groups:

$$\bar{X}_1 - \bar{X}_2 \tag{5.4}$$

The next step in the formula for the two-group t-test is to divide the answer you get when you subtract the two means by the standard error of the difference of the two means, and *this is a different standard error of the mean that you found for the one-group t-test because there are two means in the two-group t-test.*

The standard error of the mean when you have two groups of people is called the "standard error of the difference of the means" between the means of the two groups. This formula looks less scary when you break it down into four steps:

1. Square the standard deviation of Group 1, and divide this result by the sample size for Group 1 (n_1).
2. Square the standard deviation of Group 2, and divide this result by the sample size for Group 2 (n_2).
3. Add the results of the above two steps to get a total score.
4. *Take the square root of this total score* to find the standard error of the difference of the means between the two groups, $S_{\bar{X}_1 - \bar{X}_2} = \sqrt{\frac{S_1^2}{n_1} + \frac{S_2^2}{n_2}}$

This last step is the one that gives students the most difficulty when they are finding this standard error using their calculator because they are in such a hurry to get to the answer that they forget to carry the square root sign down to the last step and thus get a larger number than they should for the standard error.

5.2.1   An Example of Formula #1 for the Two-Group t-Test

Now, let us use Formula #1 in a situation in which both groups have a sample size greater than 30 people.

Suppose that you have been hired by PepsiCo to do a taste test with teenage boys (ages 13–18) to determine if they like the taste of Pepsi the same as the taste of Coke. The boys are not told the brand name of the soft drink that they taste.

You select a group of boys in this age range and randomly assign them to one of two groups: (1) Group 1 tastes Coke, and (2) Group 2 tastes Pepsi. Each group rates the taste of their soft drink on a 100-point scale using the following scale in Fig. 5.7.

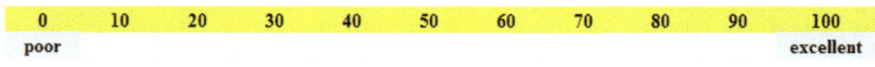

0	10	20	30	40	50	60	70	80	90	100
poor										excellent

Fig. 5.7 Example of a rating scale for a soft drink taste test (practical example)

Suppose you collect these ratings and determine (using your new Excel skills) that the 52 boys in the Coke group had a mean rating of 55 with a standard deviation of 7, while the 57 boys in the Pepsi group had a mean rating of 64 with a standard deviation of 13.

Note that the two-group t-test does not require that both groups have the same sample size. This is another way of saying that the two-group *t*-test is "robust" (a fancy term that statisticians like to use).

Your data then produce the following table in Fig. 5.8.

	A	B	C	D	E
1					
2					
3		Group	n	Mean	STDEV
4		1 Coke	52	55	7
5		2 Pepsi	57	64	13

Fig. 5.8 Worksheet data for soft drink taste test

Create an Excel spreadsheet, and enter the following information:

B3:   Group
B4:   1 Coke
B5:   2 Pepsi
C3:   *n*
D3:   Mean

E3:   STDEV
C4:   52
D4:   55
E4:   7
C5:   57
D5:   64
E5:   13

Now, widen column B so that it is twice as wide as column A, and center the six
numbers and their labels in your table (see Fig. 5.9).

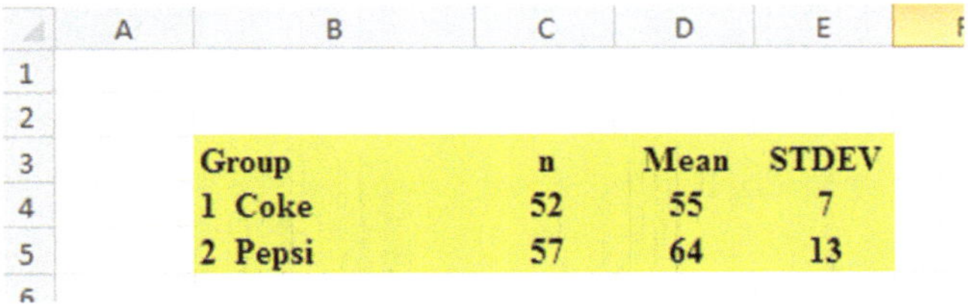

	A	B	C	D	E	F
1						
2						
3		Group	n	Mean	STDEV	
4		1 Coke	52	55	7	
5		2 Pepsi	57	64	13	
6						

Fig. 5.9 Results of widening column B and centering the numbers in the cells

B8:   Null hypothesis
B10:   Research hypothesis

Since both groups have a sample size greater than 30, you need to use Formula
#1 for the t-test for the difference of the means of the two groups.

Let us "break this formula down into pieces" to reduce the chance of making
a mistake.

B13:   STDEV1 squared/n1 (note that you square the standard deviation of Group
       1, and then divide the result by the sample size of Group 1)
B16:   STDEV2 squared/n2
B19:   D13 + D16
B22:   s.e.
B25:   critical t
B28: t-test
B31:   Result
B36:   Conclusion (see Fig. 5.10)

Group	n	Mean	STDEV
1 Coke	52	55	7
2 Pepsi	57	64	13

Null hypothesis:

Research hypothesis:

STDEV1 squared / n1

STDEV2 squared / n2

D13 + D16

s.e.

critical t

t-test

Result:

Conclusion:

Fig. 5.10 Formula labels for the two-group *t*-test

You now need to compute the values of the above formulas in the following cells:

D13:   the result of the formula needed to compute cell B13 (use two decimals)
D16:   the result of the formula needed to compute cell B16 (use two decimals)
D19:   the result of the formula needed to compute cell B19 (use two decimals)
D22:   =SQRT(D19) (use two decimals)

This formula should give you a standard error (s.e.) of 1.98.

D25: 1.96
      (Since df $= n_1 + n_2 - 2$, this gives df $= 109 - 2 = 107$, and the critical t is, therefore, 1.96 in Appendix E).

D28: =(D4–D5)/D22 (use two decimals)

This formula should give you a value for the t-test of: -4.55.

Next, check to see if you have rounded off all figures in D13: D28 to two decimal places (see Fig. 5.11).

Group	n	Mean	STDEV
1 Coke	52	55	7
2 Pepsi	57	64	13

Null hypothesis:

Research hypothesis:

STDEV1 squared / n1	0.94
STDEV2 squared / n2	2.96
D13 + D16	3.91
s.e.	1.98
critical t	1.96
t-test	-4.55

Result:

Conclusion:

Fig. 5.11 Results of the t-test formula for the soft drink taste test

Now, write the following sentence in D31 to D34 to summarize the result of the study:

D31:   Since the absolute value of −4.55
D32:   is greater than the critical *t* of
D33:   1.96, we reject the null hypothesis
D34:   and accept the research hypothesis

Finally, write the following sentence in D36 to D38 to summarize the conclusion of the study in plain English:

D36:   Teenage boys rated the taste of
D37:   Pepsi as significantly better than
D38:   the taste of Coke (64 vs. 55)

Save your file as: COKE4.

Important note: *You are probably wondering why we entered both the result and the conclusion in separate cells instead of in just one cell. This is because if you enter them in one cell, you will be very disappointed when you print out your final spreadsheet because one of two things will happen that you will not like: (1) if you print the spreadsheet to fit onto only one page, the result and the conclusion will force the entire spreadsheet to be printed in such small font size that you will be unable to read it, or (2) if you do not print the final spreadsheet to fit onto one page, both the result and the conclusion will "dribble over" onto a second page instead of fitting the entire spreadsheet onto one page. In either case, your spreadsheet will not have a "professional look."*

Print this file so that it fits onto one page, and write by hand the null hypothesis and the research hypothesis on your printout.

The final spreadsheet appears in Fig. 5.12.

Now, let us use the second formula for the two-group *t*-test which we use whenever either one group or both groups have less than 30 people in them.

Objective: To use Formula #2 for the two-group *t*-test when one or both groups have less than 30 people in them

---

Now, let us look at the case when one or both groups have a sample size less than 30 people in them.

Group	n	Mean	STDEV
1 Coke	52	55	7
2 Pepsi	57	64	13

Null hypothesis:	$\mu_1 = \mu_2$
Research hypothesis:	$\mu_1 \neq \mu_2$

STDEV1 squared / n1	0.94
STDEV2 squared / n2	2.96
D13 + D16	3.91
s.e.	1.98
critical t	1.96
t-test	-4.55

Result:	Since ther absolute value of - 4.55 is greater than the critical t of 1.96, we reject the null hypothesis and accept the research hypothesis.
Conclusion:	Teenage boys rated the taste of Pepsi as significantly better than the taste of Coke (64 vs. 55)

Fig. 5.12 Final worksheet for the Coke vs. Pepsi taste test

5.3 Formula #2: One or Both Groups Have Less Than 30 People in Them

Suppose that a school principal wanted to try out a new method of teaching reading to fourth graders and to compare it to the traditional method of teaching reading in her school. She obtained permission from the school superintendent to do a "pilot test" using two teachers from her school that she considered to be of comparable teaching ability, education, degrees earned, and years of teaching experience. The method of teaching reading that has been used by this school was called the "traditional approach," while the new method of teaching reading was called the "experimental approach." Suppose, further, that these two classes of students had very similar grade equivalent scores in reading comprehension at the end of the third grade.

Each of these teachers used just their assigned method of teaching reading, and both teachers taught reading the same amount of time during the school year. At the end of the year, both classes took the Iowa Test of Basic Skills (ITBS), and the grade equivalent scores (GE) in reading comprehension were recorded to compare the reading achievement scores of these two classes. For example, a GE score of 4.8 would mean that this pupil was reading at the same developmental level that was typical of a fourth grade pupil in the 8 month of the school year.

Suppose that you have been asked to analyze the data from the reading comprehension test scores and to compare the scores of the two classes of pupils using the two-group *t*-test for independent samples. The hypothetical data is given in Fig. 5.13.

	A	B	C	D	E	F
1		Grade 4 Iowa Tests of Basic Skills: Reading Comprehension				
2		Grade Equivalent (GE) scores				
3						
4		Group	n	Mean	STDEV	
5		1 Traditional approach	26	4.8	0.6	
6		2 Experimental approach	22	5.1	0.4	
7						

Fig. 5.13 Worksheet data for reading comprehension scores (practical example)

Null hypothesis: $\mu_1 = \mu_2$
Research hypothesis: $\mu_1 \neq \mu_2$

Note: Since both groups have a sample size less than 30 people, you need to use Formula #2 in the following steps:

Create an Excel spreadsheet, and enter the following information:

B1:   Grade 4 Iowa Tests of Basic Skills: Reading Comprehension
B2:   Grade Equivalent (GE) scores
B4:   Group
B5:   1 Traditional approach
B6:   2 Experimental approach
C4:   *n*
D4:   Mean
E4:   STDEV

Now, widen column B so that it is three times as wide as column A.

To do this, click on B at the top left of your spreadsheet to highlight all of the cells in column B. Then, move the mouse pointer to the right end of the B cell until you get a "cross" sign; then, click on this cross sign and drag the sign to the right until you can read all of the words on your screen. Then, stop clicking!

C5: 26
D5: 4.8
E5: 0.6
C6: 22
D6: 5.1
E6: 0.4

Next, *center the information in cells C4 to E6* by highlighting these cells and then using this step:

Click on the bottom line, second from the left icon, under "alignment" at the top center of Home

B9:   Null hypothesis
B11:   Research hypothesis (see Fig. 5.14)

Grade 4 Iowa Tests of Basic Skills: Reading Comprehension
Grade Equivalent (GE) scores

Group	n	Mean	STDEV
1 Traditional approach	26	4.8	0.6
2 Experimental approach	22	5.1	0.4

Null hypothesis:

Research hypothesis:

Fig. 5.14 Reading comprehension worksheet data for hypothesis testing

Since both groups have a sample size less than 30, you need to use Formula #2 for the *t*-test for the difference of the means of two independent samples.

Formula #2 for the two-group *t*-test is the following:

$$t = \frac{\bar{X}_1 - \bar{X}_2}{S_{\bar{X}_1 - \bar{X}_2}} \qquad (5.2)$$

$$where\ S_{\bar{X}_1 - \bar{X}_2} = \sqrt{\frac{(n_1 - 1)S_1{}^2 + (n_2 - 1)S_2{}^2}{n_1 + n_2 - 2} \left(\frac{1}{n_1} + \frac{1}{n_2}\right)} \qquad (5.5)$$

and where degrees of freedom $= df = n_1 + n_2 - 2 \qquad (5.1)$

This formula is complicated, and so, it will reduce your chance of making a mistake in writing it if you "break it down into pieces" instead of trying to write the formula as one cell entry.

Now, enter these words on your spreadsheet:

B14: $(n_1 - 1) \times$ STDEV1 squared
B17: $(n_2 - 1) \times$ STDEV2 squared
B20: $n_1 + n_2 - 2$
B23: $1/n_1 + 1/n_2$
B26:   s.e.
B29:   critical *t*
B32:   *t*-test
B35:   Result
B40:   Conclusion (see Fig. 5.15)

You now need to compute the values of the above formulas in the following cells:

E14:   the result of the formula needed to compute cell B14 (use two decimals)
E17:   the result of the formula needed to compute cell B17 (use two decimals)
E20:   the result of the formula needed to compute cell B20
E23:   the result of the formula needed to compute cell B23 (use two decimals)
E26:   =SQRT(((E14 + E17)/E20)*E23)

Note the three open parentheses after SQRT and the three closed parentheses on the right side of this formula. You need three open parentheses and three closed parentheses in this formula or the formula will not work correctly.

The above formula gives a standard error of the difference of the means equal to 0.15 (two decimals).

E29:   enter the critical *t* value from the *t*-table in Appendix E in this cell using
       df $= n_1 + n_2 - 2$ to find the critical *t* value
E32:   =(D5−D6)/E26

Note that you need an open parenthesis *before D5* and a closed parenthesis *after D6* so that this answer of −0.30 is *THEN* divided by the standard error of the difference of the means of 0.15 to give a *t*-test value of −2.00 (note the minus sign here). Use two decimal places for the *t*-test result (see Fig. 5.16).

Now write the following sentence in D35 to D38 to summarize the *result* of the study:

D35:   Since the absolute value of t
D36:   of −2.00 is greater than the critical t
D37:   of 1.96, we reject the null hypothesis and
D38:   accept the research hypothesis

Grade 4 Iowa Tests of Basic Skills: Reading Comprehension
Grade Equivalent (GE) scores

Group	n	Mean	STDEV
1 Traditional approach	26	4.8	0.6
2 Experimental approach	22	5.1	0.4

Null hypothesis:

Research hypothesis:

(n1 - 1) x STDEV1 squared

(n2 - 1) x STDEV2 squared

n1 + n2 - 2

1/n1 + 1/n2

s.e.

critical t

t-test

Result:

Conclusion:

Fig. 5.15 Reading comprehension formula labels for two-group *t*-test

Finally, write the following sentence in D40 to D42 to summarize the *conclusion* of the study:

D40: The experimental group had significantly
D41: higher grade equivalent (GE) scores than
D42: the traditional group (5.1 vs. 4.8).

Save your file as: Reading3.

Print the final spreadsheet so that it fits onto one page.
Write the null hypothesis and the research hypothesis by hand on your printout.
The final spreadsheet appears in Fig. 5.17.

Grade 4 Iowa Tests of Basic Skills: Reading Comprehension
Grade Equivalent (GE) scores

Group	n	Mean	STDEV
1 Traditional approach	26	4.8	0.6
2 Experimental approach	22	5.1	0.4

Null hypothesis:

Research hypothesis:

(n1 - 1) x STDEV1 squared	9.00
(n2 - 1) x STDEV2 squared	3.36
n1 + n2 - 2	46
1/n1 + 1/n2	0.08
s.e.	0.15
critical t (df = 46)	1.96
t-test	-2.00

Result:

Conclusion:

Fig. 5.16 Reading comprehension two-group *t*-test formula results

5.4   End-of-Chapter Practice Problems

1. Suppose that a school superintendent wanted to try out a new method of teaching mathematics to eighth graders and that he has decided to ask the math teachers in one middle school to use the traditional approach and the math teachers in another school to use the experimental approach. Assume that the math teachers in both schools have very similar teaching ability, education, years of teaching experience, and credits earned, and that the students in both schools had very similar math achievement scores at the end of the seventh grade.

Grade 4 Iowa Tests of Basic Skills: Reading Comprehension
Grade Equivalent (GE) scores

Group	n	Mean	STDEV
1 Traditional approach	26	4.8	0.6
2 Experimental approach	22	5.1	0.4

Null hypothesis:	$\mu_1 = \mu_2$
Research hypothesis:	$\mu_1 \neq \mu_2$

(n1 - 1) x STDEV1 squared	9.00
(n2 - 1) x STDEV2 squared	3.36
n1 + n2 - 2	46
1/n1 + 1/n2	0.08
s.e.	0.15
critical t	1.96
t-test	-2.00

Result:	Since the absolute value of t of - 2.00 is greater than the critical t of 1.96, we reject the null hypothesis and accept the research hypothesis.
Conclusion:	The experimental group had significantly higher grade equivalent (GE) scores than the traditional group (5.1 vs. 4.8).

Fig. 5.17 Reading comprehension scores final spreadsheet

Suppose that the Iowa Test of Basic Skills (ITBS) was used at the end of the eighth grade and that grade equivalent (GE) scores in the math problem solving and data interpretation subtest were used to compare the two teaching methods. The traditional approach was used on 124 pupils who had a mean GE score of 8.7 and a standard deviation of 0.8. The experimental approach was used on 135 pupils who had a mean GE score of 8.9 and a standard deviation of 0.7.

(a) State the null hypothesis and the research hypothesis on an Excel spreadsheet.
(b) Find the standard error of the difference between the means using Excel.

(c) Find the critical *t* value using Appendix E, and enter it on your spreadsheet.

(d) Perform a *t*-test on these data using Excel. What is the value of *t* that you obtain?

   Use three decimal places for all figures in the formula section of your spreadsheet.

(e) State your result on your spreadsheet.

(f) State your conclusion in plain English on your spreadsheet.

(g) Save the file as: Math14.

2. Massachusetts Mutual Financial Group (2010) placed a full-page color ad in *The Wall Street Journal* in which it used a male model hugging a 2-year old daughter. The ad had the headline and subheadline:

WHAT IS THE SIGN OF A GOOD DECISION?

   It's knowing your life insurance can help provide income for retirement. And peace of mind until you get there.

   Since the majority of the subscribers to *The Wall Street Journal* are men, an interesting research question would be the following:

   Research question: "Does a male model in a magazine ad affect adult men's or adult women's willingness to learn more about how life insurance can provide income for retirement?"

   Suppose that you have shown one group of adult males (ages 25–39) and one group of adult females (ages 25–39) a mock-up of an ad such that both groups saw the ad with a male model. The ads were identical in copy format. The two groups were kept separate during the experiment and could not interact with one another.

   At the end of a 1-hour discussion of the mock-up ad, the respondents were asked the question given in Fig. 5.18.

The resulting hypothetical data for this question appear in Fig. 5.19.

(a) On your Excel spreadsheet, write the null hypothesis and the research hypothesis.

(b) Create a table that summarizes these data on your spreadsheet and use Excel to find the sample sizes, the means, and the standard deviations of the two groups in this table.

(c) Use Excel to find the standard error of the difference of the means.

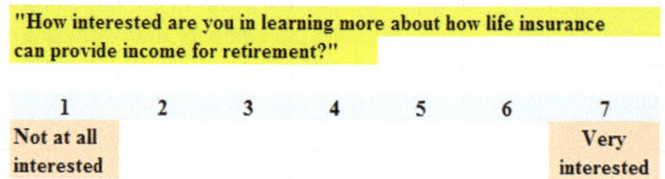

Fig. 5.18 Rating scale item for a magazine ad interest indicator (practical example)

Magazine ad: Male model	
Men	**Women**
5	3
6	4
4	6
7	5
5	2
6	3
5	1
4	3
3	2
6	4
7	3
5	5
6	6
4	3
7	4
5	2
4	5
6	3
3	4
7	5
5	4
6	3
2	2
6	4
1	3
7	5
6	1
5	3
4	2
6	3
5	2
7	5
	3
	4

Fig. 5.19 Worksheet data for Chap. 5: practice problem #2

(d) Use Excel to perform a two-group t-test. What is the value of t that you obtain (use two decimal places)?
(e) On your spreadsheet, type the *critical value of t* using the t-table in Appendix E.
(f) Type your *result* on the test on your spreadsheet.

(g) Type your *conclusion in plain English* on your spreadsheet.
(h) Save the file as: lifeinsur12.

3. Suppose that a school superintendent wanted to try out a new method of teaching vocabulary to third graders and that he has decided to ask one teacher in one elementary school to use the traditional approach to teaching vocabulary, and another teacher in a different school to use an experimental approach to teaching vocabulary. Assume that the teachers have very similar teaching ability, education, years of teaching experience, and credits earned and that the students in both schools had very similar vocabulary test scores at the end of the second grade.

Suppose that the Iowa Test of Basic Skills (ITBS) was used at the end of the third grade and that grade equivalent (GE) scores in the vocabulary subtest were used to compare the two teaching methods.

The traditional approach was used on 18 pupils who had a mean GE score of 3.7 and a standard deviation of 0.8. The experimental approach was used on 19 pupils who had a mean GE score of 4.3 and a standard deviation of 0.7.

(a) State the null hypothesis and the research hypothesis on an Excel spreadsheet.
(b) Find the standard error of the difference between the means using Excel.
(c) Find the critical *t* value using Appendix E, and enter it on your spreadsheet.
(d) Perform a *t*-test on these data using Excel. What is the value of *t* that you obtain?
(e) State your result on your spreadsheet.
(f) State your conclusion in plain English on your spreadsheet.
(g) Save the file as: Vocabulary4.

References

Keller, G. Statistics for Management and Economics (8th ed.). Mason, OH: South Western Cengage Learning, 2009.

Zikmund, W.G. and Babin, B.J. Exploring Marketing Research (10th ed.). Mason, OH: South-Western Cengage Learning, 2010.

Mass Mutual Financial Group. What is the Sign of a Good Decision? (Advertisement) *The Wall Street Journal*, September 29, 2010, p. A22.

Chapter 6
Correlation and Simple Linear Regression

There are many different types of "correlation coefficients," but the one we will use in this book is the Pearson product–moment correlation which we will call r.

6.1 What Is a "Correlation"?

Basically, a correlation is a number between -1 and $+1$ that summarizes the relationship between two variables, which we will call X and Y.

A correlation can be either positive or negative. *A positive correlation means that as X increases, Y increases. A negative correlation means that as X increases, Y decreases.* In statistics books, this part of the relationship is called the *direction* of the relationship (i.e., it is either positive or negative).

The correlation also tells us the *magnitude* of the relationship between X and Y. As the correlation approaches closer to $+1$, we say that the relationship is *strong and positive*.

As the correlation approaches closer to -1, we say that the relationship is *strong and negative*.

A zero correlation means that there is no relationship between X and Y. This means that neither X nor Y can be used as a predictor of the other.

A good way to understand what a correlation means is to see a "picture" of the scatterplot of points produced in a chart by the data points. Let us suppose that you want to know if variable X can be used to predict variable Y. We will place *the predictor variable X on the x-axis* (the horizontal axis of a chart) and *the criterion variable Y on the y-axis* (the vertical axis of a chart). Suppose, further, that you have collected data given in the scatterplots below (see Figs. 6.1, 6.2, 6.3, 6.4, 6.5, 6.6).

Figure 6.1 shows the scatterplot for a perfect positive correlation of $r = +1.0$. This means that you can perfectly predict each Y-value from each X-value because the data points move "upward and to the right" along a perfectly fitting straight line (see Fig. 6.1).

T. Quirk, *Excel 2010 for Educational and Psychological Statistics:*
A Guide to Solving Practical Problems, DOI 10.1007/978-1-4614-2071-2_6,
© Springer Science+Business Media, LLC 2012

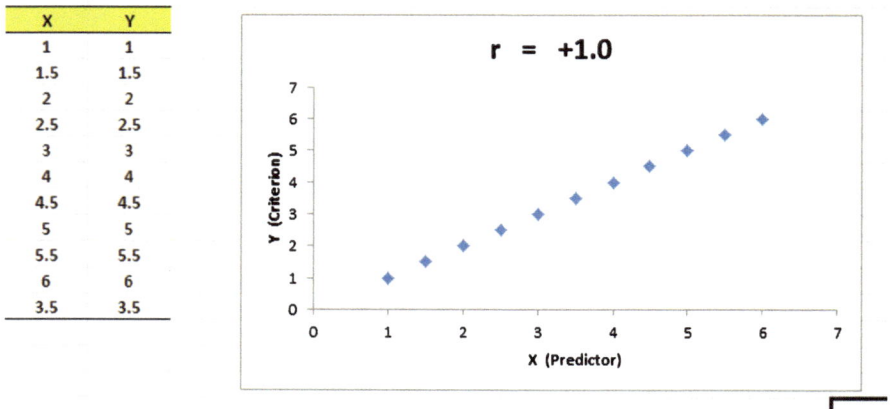

X	Y
1	1
1.5	1.5
2	2
2.5	2.5
3	3
4	4
4.5	4.5
5	5
5.5	5.5
6	6
3.5	3.5

Fig. 6.1 Example of a scatterplot for a perfect, positive correlation ($r = +1.0$)

Figure 6.2 shows the scatterplot for a moderately positive correlation of $r = +.53$. This means that each X-value can predict each Y-value moderately well because you can draw a picture of a "football" around the outside of the data points that move upward and to the right, but not along a straight line (see Fig. 6.2).

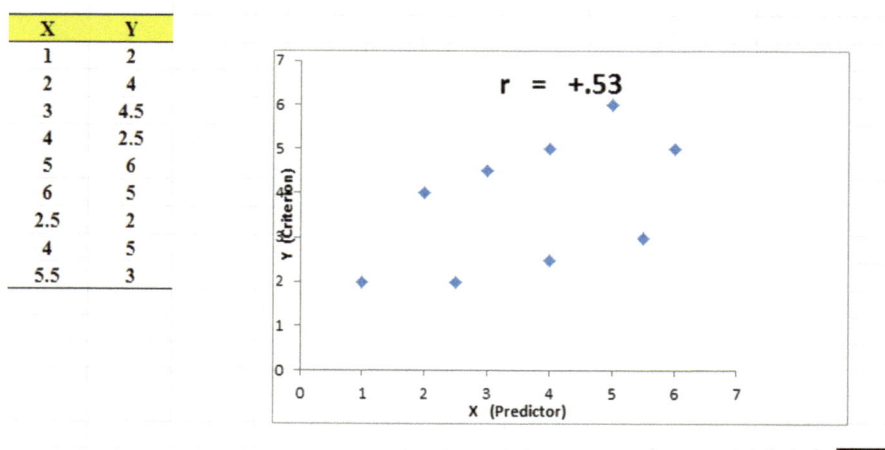

X	Y
1	2
2	4
3	4.5
4	2.5
5	6
6	5
2.5	2
4	5
5.5	3

Fig. 6.2 Example of a scatterplot for a moderate, positive correlation ($r = +.53$)

Figure 6.3 shows the scatterplot for a low, positive correlation of $r = +.23$. This means that each X-value is a poor predictor of each Y-value because the "picture" you could draw around the outside of the data points approaches a circle in shape (see Fig. 6.3).

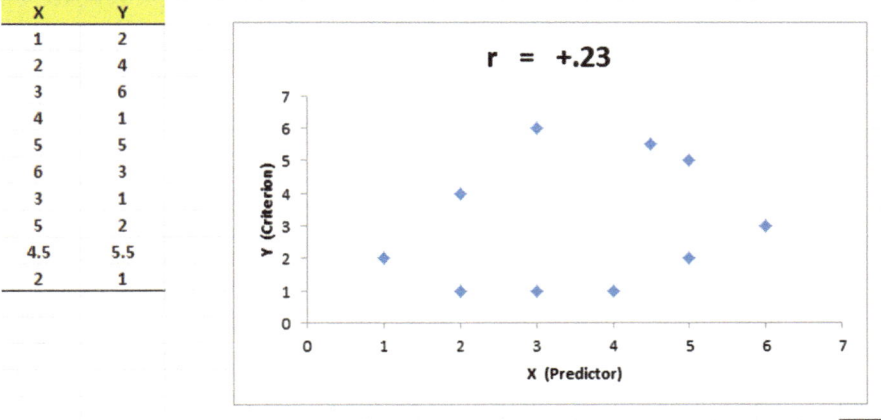

X	Y
1	2
2	4
3	6
4	1
5	5
6	3
3	1
5	2
4.5	5.5
2	1

Fig. 6.3 Example of a scatterplot for a low, positive correlation ($r = +.23$)

We have not shown a figure of a zero correlation because it is easy to imagine what it looks like as a scatterplot. A zero correlation of $r = .00$ means that there is no relationship between X and Y, and the "picture" drawn around the data points would be a perfect circle in shape, indicating that you cannot use X to predict Y because these two variables are not correlated with one another.

Figure 6.4 shows the scatterplot for a low, negative correlation of $r = -.22$ which means that each X is a poor predictor of Y in an inverse relationship, meaning that as X increases, Y decreases (see Fig. 6.4). In this case, it is a negative correlation because the "football" you could draw around the data points slopes down and to the right.

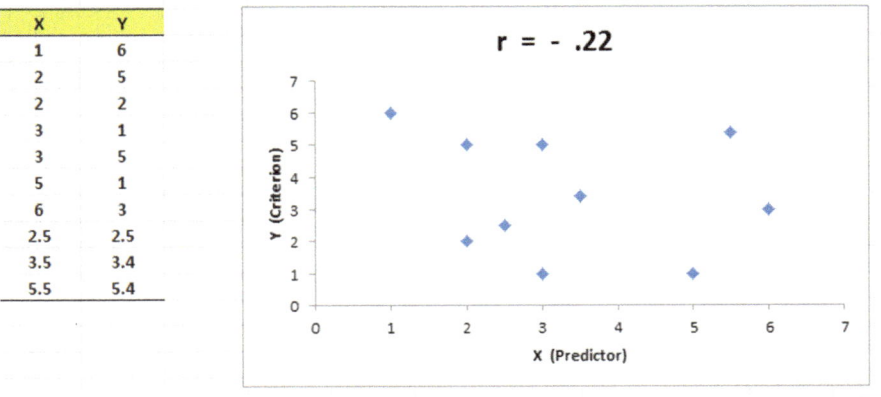

X	Y
1	6
2	5
2	2
3	1
3	5
5	1
6	3
2.5	2.5
3.5	3.4
5.5	5.4

Fig. 6.4 Example of a scatterplot for a low, negative correlation ($r = -.22$)

Figure 6.5 shows the scatterplot for a moderate, negative correlation of $r = -.39$, which means that X is a moderately good predictor of Y, although there is an inverse relationship between X and Y (i.e., as X increases, Y decreases; see Fig. 6.5). In this case, it is a negative correlation because the "football" you could draw around the data points slopes down and to the right.

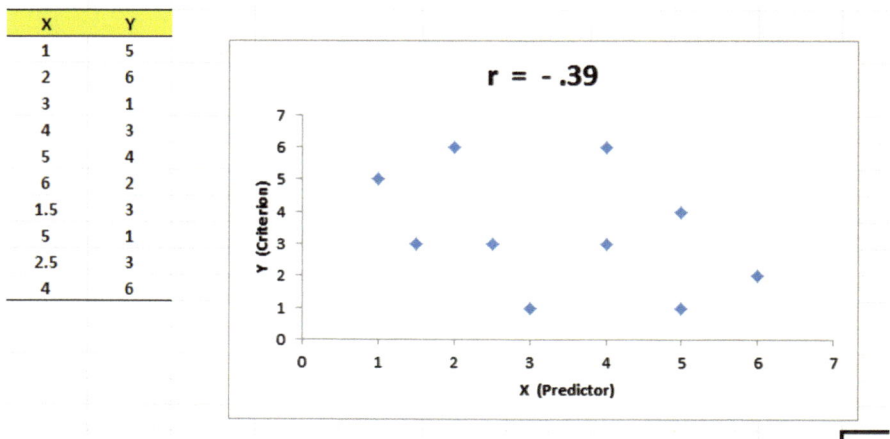

X	Y
1	5
2	6
3	1
4	3
5	4
6	2
1.5	3
5	1
2.5	3
4	6

Fig. 6.5 Example of a scatterplot for a moderate, negative correlation ($r = -.39$)

Figure 6.6 shows a perfect negative correlation of $r = -1.0$, which means that X is a perfect predictor of Y, although in an inverse relationship such that as X increases, Y decreases. The data points fit perfectly along a downward-sloping straight line (see Fig. 6.6).

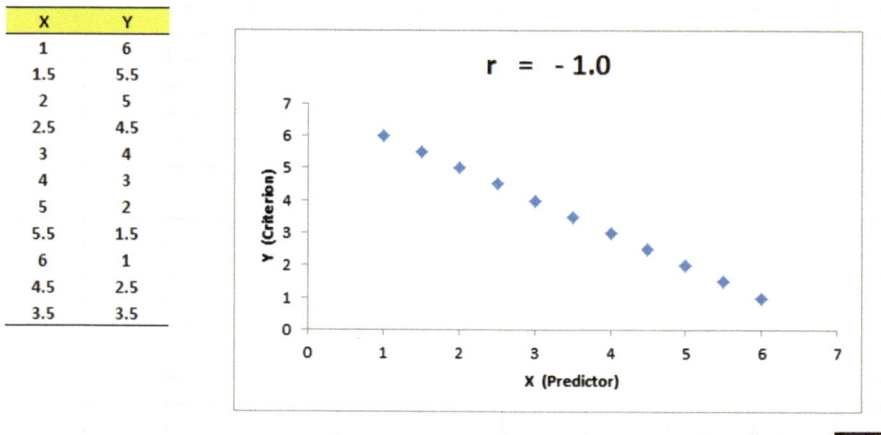

X	Y
1	6
1.5	5.5
2	5
2.5	4.5
3	4
4	3
5	2
5.5	1.5
6	1
4.5	2.5
3.5	3.5

Fig. 6.6 Example of a scatterplot for a perfect, negative correlation ($r = -1.0$)

Let us explain the formula for computing the correlation r so that you can understand where the number summarizing the correlation came from.

In order to help you to understand *where* the correlation number that ranges from -1.0 to $+1.0$ comes from, we will walk you through the steps involved to use the formula as if you were using a pocket calculator. This is the one time in this book that we will ask you to use your pocket calculator to find a correlation, but knowing how the correlation is computed step by step will give you the opportunity to understand *how* the formula works in practice.

To do that, let us create a situation in which you need to find the correlation between two variables.

Suppose that you wanted to find out if there was a relationship between high school grade-point average (HSGPA) and freshman GPA (FRGPA) at a liberal arts college. You have decided to call HSGPA the X-variable (i.e., the predictor variable) and FRGPA as the Y-variable (i.e., the criterion variable) in your analysis. To test your Excel skills, you take a random sample of freshmen at the end of their freshman year and record their GPA. The hypothetical data for eight students appear in Fig. 6.7. (*Note: We are using only one decimal place for these GPAs in this example to simplify the mathematical computations*).

	A	B	C	
1				
2		X	Y	
3	Student	High School GPA	FROSH GPA	
4	1	2.8	2.9	
5	2	2.5	2.8	
6	3	3.1	2.8	
7	4	3.5	3.2	
8	5	2.4	2.6	
9	6	2.6	2.3	
10	7	2.4	2.1	
11	8	3.6	3.2	
12				
13	n	8	8	
14	MEAN	2.86	2.74	
15	STDEV	0.48	0.39	
16				

Fig. 6.7 Worksheet data for high school GPA and Frosh GPA (practical example)

Notice also that we have used Excel to find the sample size for both variables, X and Y, and the MEAN and STDEV of both variables. (You can practice your Excel skills by seeing if you get these same results when you create an Excel spreadsheet for these data).

Now, let us use the above table to compute the correlation r between HSGPA and FRGPA using your pocket calculator.

6.1.1   Understanding the Formula for Computing a Correlation

Objective: To understand the formula for computing the correlation r

The formula for computing the correlation r is as follows:

$$r = \frac{\frac{1}{n-1} \Sigma (X - \bar{X})(Y - \bar{Y})}{S_x S_y} \tag{6.1}$$

This formula looks daunting at first glance, but let us "break it down into its steps" to understand how to compute the correlation r.

6.1.2   Understanding the Nine Steps for Computing a Correlation, r

Objective: To understand the nine steps of computing a correlation r

The nine steps are as follows:

Step	Computation	Result
1	Find the sample size (n) by noting the number of students.	8
2	Divide the number 1 by the sample size minus 1 (i.e., 1/7).	0.14286
3	*For each student*, take the HSGPA and subtract the mean HSGPA for the eight students and call this $X - \bar{X}$ (e.g., for student # 6, this would be 2.6–2.86).	− 0.26
	Note: With your calculator, this difference is −0.26, but when Excel uses 16 decimal places for every computation, this result could be slightly different for each student.	
4	*For each student*, take the FRGPA and subtract the mean FRGPA for the eight students and call this $Y - \bar{Y}$ (e.g., for student # 6, this would be 2.3–2.74).	−0.44
5	Then, *for each student*, multiply $(X - \bar{X})$ times $(Y - \bar{Y})$ (e.g., for student # 6 this would be $(-0.26) \times (-0.44)$).	+0.1144
6	Add the results of $(X - \bar{X})$ times $(Y - \bar{Y})$ for the eight students.	+1.09

Steps 1–6 would produce the Excel table given in Fig. 6.8.

	A	B	C	D	E	F	G
1							
2		X	Y				
3	**Student**	**High School GPA**	**FROSH GPA**	$X - \bar{X}$	$Y - \bar{Y}$	$(X - \bar{X})(Y - \bar{Y})$	
4	1	2.8	2.9	-0.06	0.16	-0.01	
5	2	2.5	2.8	-0.36	0.06	-0.02	
6	3	3.1	2.8	0.24	0.06	0.01	
7	4	3.5	3.2	0.64	0.46	0.29	
8	5	2.4	2.6	-0.46	-0.14	0.06	
9	6	2.6	2.3	-0.26	-0.44	0.11	
10	7	2.4	2.1	-0.46	-0.64	0.29	
11	8	3.6	3.2	0.74	0.46	0.34	
12						- - - - - - -	
13	**n**	8	8		Total	1.09	
14	**MEAN**	2.86	2.74				
15	**STDEV**	0.48	0.39				
16							

Fig. 6.8 Worksheet for computing the correlation, r

Notice that when Excel multiplies a minus number by a minus number, the result is a plus number (e.g., for student #7 $(-0.46) \times (-0.64) = +0.29$. And when Excel multiplies a minus number by a plus number, the result is a negative number (e.g., for student #1 $(-0.06) \times (+0.16) = -0.01$.

Note: Excel computes all computation to 16 decimal places. So, when you check your work with a calculator, you frequently get a slightly different answer than Excel's answer.

For example, when you compute above

$$(X - \bar{X}) \times (Y - \bar{Y}) \text{ for student } \#2, \text{ your calculator gives} \qquad (6.2)$$

$$(0.36) \times (+0.06) = -0.0216.$$

As you can see from the table, Excel's answer is -0.02, which is really *more accurate* because Excel uses 16 decimal places for every number, even though only two decimal places are shown in Fig. 6.8.

You should also note that when you do Step 6, you have to be careful to add all of the positive numbers first to get $+1.10$ and then add all of the negative numbers second to get -0.03, so that when you subtract these two numbers, you get $+1.07$ as your answer to Step 6. When you do these computations using Excel, this total figure will be $+1.09$ because Excel carries every number and computation out to 16 decimal places which is much more accurate than your calculator.

Step

7	Multiply the answer for Step 2 above by the answer for Step 6 (0.14286×1.09).	0.1557
8	Multiply the STDEV of X times the STDEV of Y (0.48×0.39).	0.1872
9	Finally, divide the answer from Step 7 by the answer from Step 8 $(0.1557$ divided by $0.1872)$.	+0.83

This number of *0.83* is the correlation between HSGPA (*X*) and FRGPA (*Y*) for these eight students. The number +*0.83* means that there is a strong, positive correlation between these two variables. That is, as HSGPA increases, FRGPA increases. For a more detailed discussion of correlation, see Zikmund and Babin (2010).

You could also use the results of the above table in the formula for computing the correlation *r* in the following way:

$$\text{Correlation } R = \left[(1/(n-1)) \times \sum (X - \overline{X})(Y - \overline{Y}) \right] / (\text{STDEV}_x \times \text{STDEV}_y)$$

$$\text{Correlation } r = [(1/7) \times 1.09]/[(.48) \times (.39)]$$

$$\text{Correlation} = r = 0.83$$

When you use Excel for these computations, you obtain a slightly different correlation of +0.82 because Excel uses 16 decimal places for all numbers and computations and is, therefore, more accurate than your calculator.

Now, let us discuss how you can use Excel to find the correlation between two variables in a much simpler, and much faster, fashion than using your calculator.

6.2 Using Excel to Compute a Correlation Between Two Variables

---

Objective: To use Excel to find the correlation between two variables

---

Suppose that you have been asked to study the relationship between scores on the Law School Admission Test (LSAT) and the GPA of students at the end of their first year of Law School. The LSAT is a standardized objective measure of Law School applicants and is a required exam for all Law Schools in the USA that are approved by the American Bar Association. About 150,000 applicants take this exam every year in the USA. Because colleges differ in their standards for grades in courses, the LSAT provides a "level playing field" for all applicants by measuring their readiness for Law School in a single examination taken by all the applicants. There are three subtests of the LSAT (reading comprehension, analytical reasoning, and logical reasoning) that produce a single score that ranges between 120 and 180, with an average score about 150.

To test your Excel skills, you take a random sample of students at the end of their first year of Law School and record their GPA. The hypothetical data appear in Fig. 6.9.

	A	B	C	D	E	F
1						
2	**LAW SCHOOL ADMISSION TEST (LSAT)**					
3						
4	Is there a relationship between LSAT scores and first-year GPA in law school?					
5						
6		**LSAT score**	**First-year Law School GPA**			
7		130	2.65			
8		170	3.72			
9		140	2.85			
10		160	3.25			
11		150	2.75			
12		180	3.95			
13		130	2.35			
14		160	2.74			
15		170	3.65			
16		140	2.55			
17		160	3.72			
18		140	2.35			

Fig. 6.9 Worksheet data for LSAT scores and GPA (practical example)

You want to determine if there is a *relationship* between the LSAT scores and GPA at the end of the first year of Law School, and you decide to use a correlation to determine this relationship. Let us call the LSAT scores the predictor, X, and first-year GPA the criterion, Y.

Create an Excel spreadsheet with the following information:

A2:   LAW SCHOOL ADMISSION TEST (LSAT)
A4:   Is there a relationship between LSAT scores and first-year GPA
          in law school?
B6:   LSAT score
C6:   First-year Law School GPA
B7:   130

Next, change the width of columns B and C so that the information fits inside the cells.

Now, complete the remaining figures in the table given above so that B18 is 140 and C18 is 2.35. (Be sure to double-check your figures to make sure that they are correct!) Then, center the information in all of these cells.

A20:   *n*
A21:   mean
A22:   stdev

Next, define the "name" to the range of data from B7:B18 as: LSAT.

We discussed earlier in this book (see Sect. 1.4.4) how to "name a range of data," but here is a reminder of how to do that.

To give a "name" to a range of data,

Click on the top number in the range of data and drag the mouse down to the bottom number of the range.

For example, to give the name "LSAT" to the cells B7:B18, click on B7 and drag the pointer down to B18 so that the cells B7:B18 are highlighted on your computer screen. Then, click on:

Formulas.

Define name (top center of your screen).

LSAT (in the Name box; see Fig. 6.10).

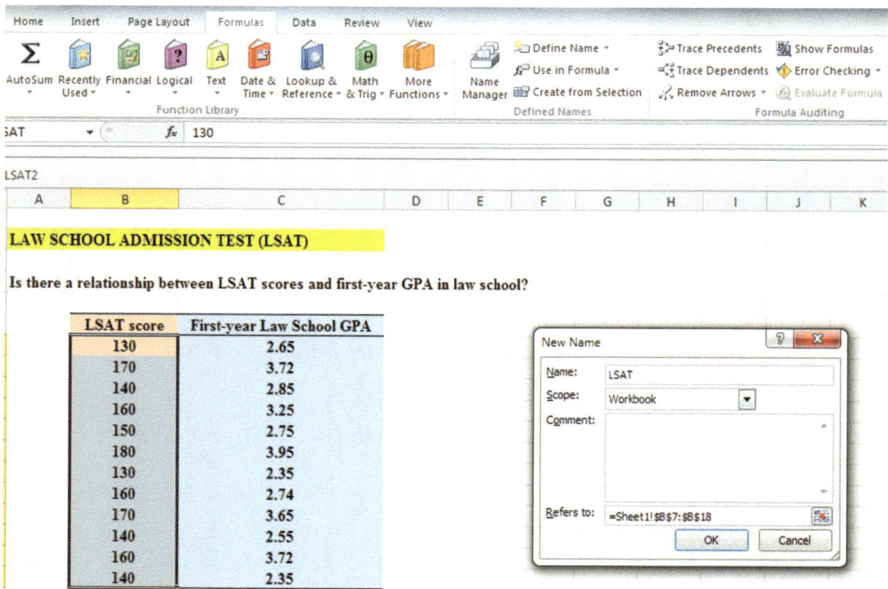

Fig. 6.10 Dialog box for naming a range of data as "LSAT"

OK

Now, repeat these steps to give the name *GPA* to C7:C18.

Finally, click on any blank cell on your spreadsheet to "deselect" cells C7:C18 on your computer screen.

Now, complete the data for these sample sizes, means, and standard deviations in columns B and C so that B22 is 16.58 and C22 is 0.58 (use two decimals for the means and standard deviations; see Fig. 6.11).

LAW SCHOOL ADMISSION TEST (LSAT)

Is there a relationship between LSAT scores and first-year GPA in law school?

LSAT score	First-year Law School GPA
130	2.65
170	3.72
140	2.85
160	3.25
150	2.75
180	3.95
130	2.35
160	2.74
170	3.65
140	2.55
160	3.72
140	2.35

n	12	12
mean	152.50	3.04
stdev	16.58	0.58

Fig. 6.11 Example of using Excel to find the sample size, mean, and STDEV

Objective: Find the correlation between LSAT scores and first-year GPA

B24:   correlation
C24:   =correl(LSAT, GPA); see Fig. 6.12

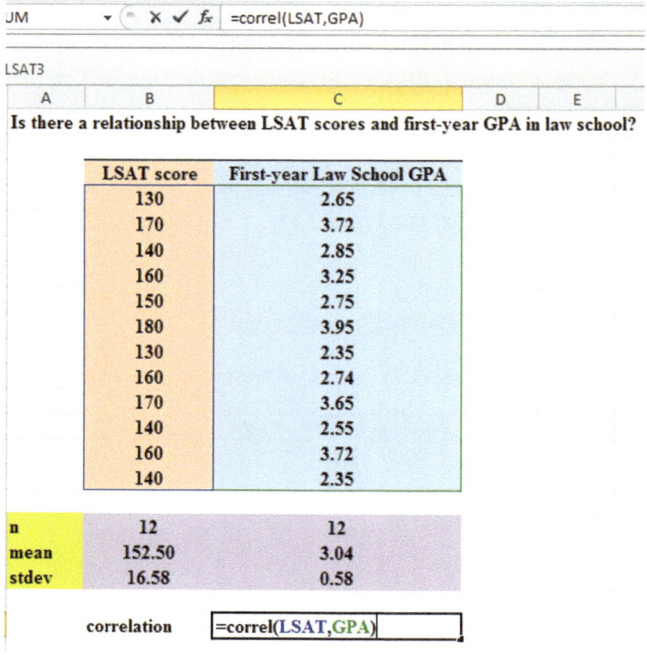

JM	▾ ⬤ ✗ ✓ ƒx	=correl(LSAT,GPA)

Fig. 6.12 Example of using Excel's =correl function to compute the correlation coefficient

Hit the Enter key to compute the correlation.
C24:   Format this cell to two decimals

Note that the equal sign in =correl(LSAT,GPA)in C24 tells Excel that you are going to use a formula in this cell.

The correlation between LSAT scores (X) and first-year GPA (Y) is +.89, a very strong positive correlation. This means that you have evidence that there is a strong relationship between these two variables. In effect, the higher the LSAT score, the higher the first-year GPA in this Law School.

Save this file as: LSAT4.

The final spreadsheet appears in Fig. 6.13.

LAW SCHOOL ADMISSION TEST (LSAT)

Is there a relationship between LSAT scores and first-year GPA in law school?

LSAT score	First-year Law School GPA
130	2.65
170	3.72
140	2.85
160	3.25
150	2.75
180	3.95
130	2.35
160	2.74
170	3.65
140	2.55
160	3.72
140	2.35

n	12	12
mean	152.50	3.04
stdev	16.58	0.58
correlation		0.89

Fig. 6.13 Final result of using the =correl function to compute the correlation coefficient

6.3   Creating a Chart and Drawing the Regression Line onto the Chart

This section deals with the concept of "linear regression." Technically, the use of a simple linear regression model (i.e., the word "simple" means that only one predictor, X, is used to predict the criterion, Y) requires that the data meet the following four assumptions if that statistical model is to be used:

1. The underlying relationship between the two variables under study (X and Y) is *linear* in the sense that a straight line, and not a curved line, can fit among the data points on the chart.
2. The errors of measurement are independent of each other (e.g., the errors from a specific time period are sometimes correlated with the errors in a previous time period).
3. The errors fit a normal distribution of Y-values at each of the X-values.
4. The variance of the errors is the same for all X-values (i.e., the variability of the Y-values is the same for both low and high values of X).

A detailed explanation of these assumptions is beyond the scope of this book, but the interested reader can find a detailed discussion of these assumptions in Levine et al. (2011, pp. 529–530).

Now, let us create a chart summarizing these data.

Important note: *Whenever you are preparing a chart, we strongly recommend that you put the predictor variable (X) on the left and the criterion variable (Y) on the right in your Excel spreadsheet, so that you do* *not get these variables backward in your Excel steps and make a mess of the problem in your computations. If you do this as a habit, you will save yourself a lot of grief.*

Let us suppose that you would like to use LSAT scores as the predictor variable and that you would like to use it to predict first-year GPA for applicants to this Law School. Since the correlation between these two variables is +.89, this shows that there is a strong, positive relationship, and that LSAT scores are a good predictor of first-year GPA.

1. Open the file that you saved earlier in this chapter, LSAT4.

6.3.1 Using Excel to Create a Chart and the Regression Line Through the Data Points

> Objective: To create a chart and the regression line summarizing the relationship between LSAT scores and first-year GPA in Law School

2. Click and drag the mouse to highlight both columns of numbers (B7:C18), *but do not highlight the labels at the top of column B and column C.*

   Highlight the data set B7:C18
   Insert (top left of screen)
   Scatter (at top of screen)
   Click on top-left chart icon under "scatter" (see Fig. 6.14)

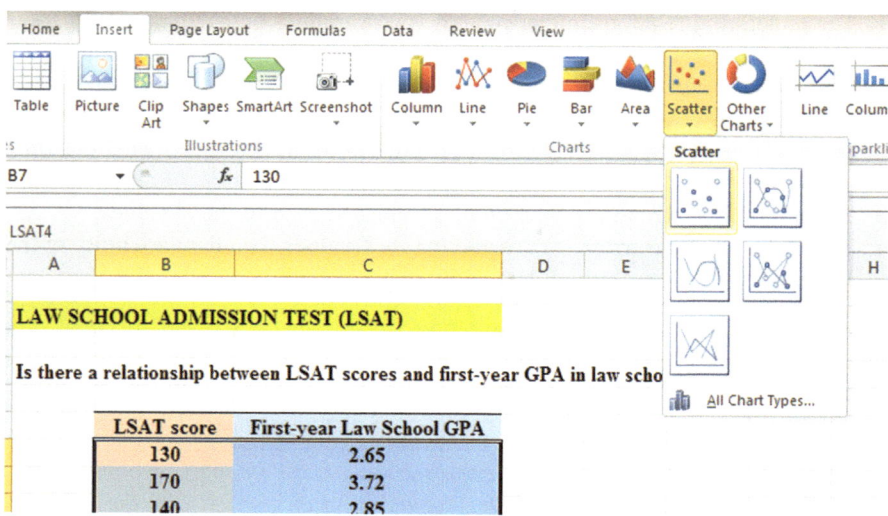

Fig. 6.14 Example of inserting a Scatter Chart into a worksheet

Layout (top right of screen under Chart Tools)
Chart title (top of screen)
Above chart (see Fig. 6.15)

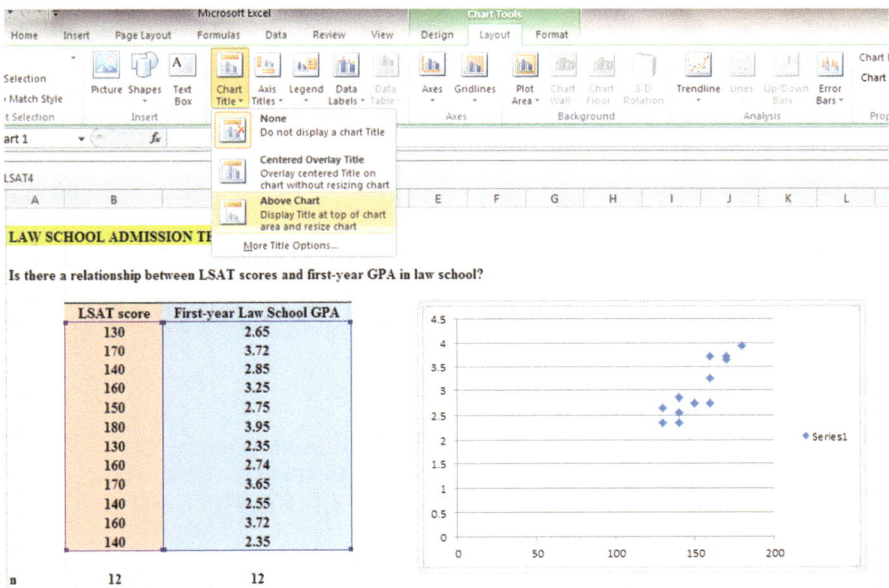

Fig. 6.15   Example of Layout/Chart Title/Above Chart commands

Enter this title in the Title box (it will appear to the right of "Chart 1 f_x" at the top of your screen):
RELATIONSHIP BETWEEN LSAT SCORES AND FIRST YEAR GPA IN LAW SCHOOL (see Fig. 6.16)

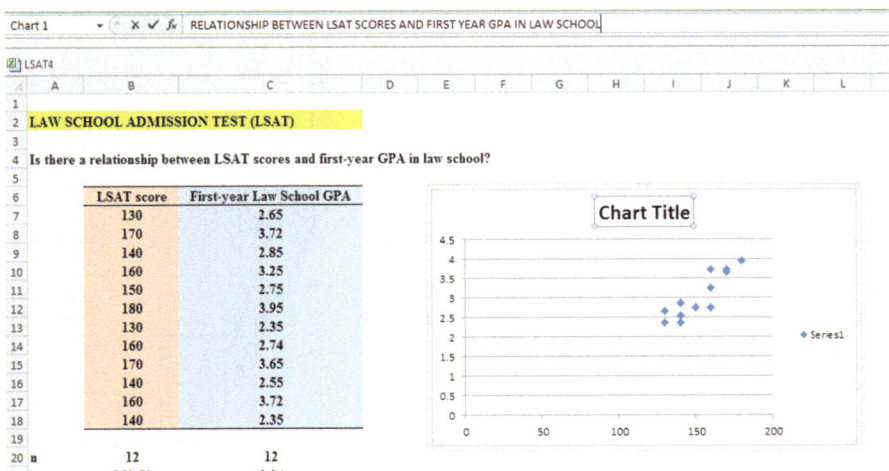

Fig. 6.16   Example of inserting the chart title above the chart

Hit the enter key to place this title above the chart.

Click on *any white space outside of the top title but inside the chart* to "deselect" this chart title.

Axis titles (at top of screen)
Primary Horizontal Axis title
Title below axis (see Fig. 6.17)

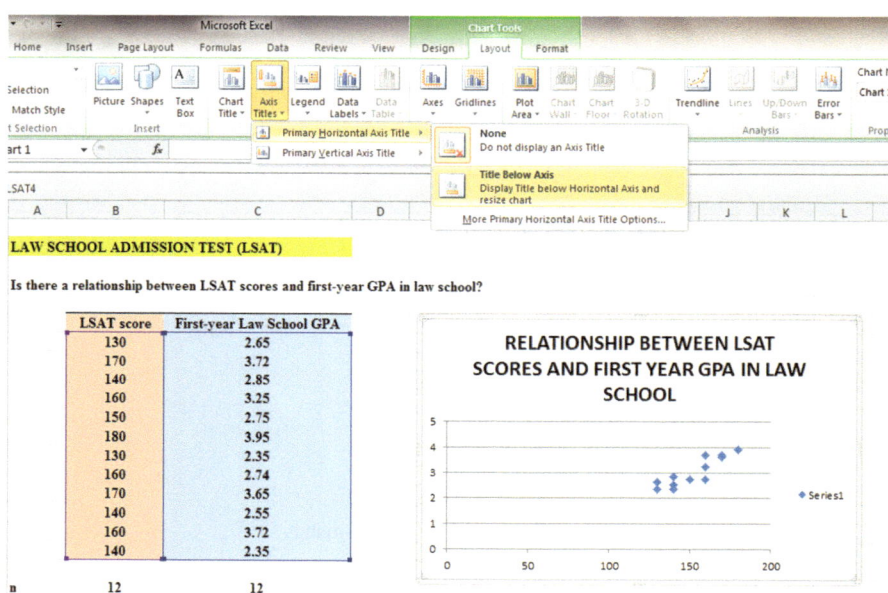

Fig. 6.17 Example of creating the *x*-axis title in a chart

Now, enter this *x*-axis title in the "Axis Title Box" at the top of your screen,

LSAT Scores.
Next, hit the enter key to place this *x*-axis title at the bottom of the chart.
Click on *any white space inside the chart but outside of this x-axis title* to "deselect" the *x*-axis title.
Axis Titles (top center of screen)
Primary Vertical Axis Title
Rotated title
Enter this *y*-axis title in the Axis Title Box at the top of your screen,
First-year GPA in Law School.
Next, hit the enter key to place this *y*-axis title along the *y*-axis.
Then, click on *any white space inside the chart but outside this y-axis title* to "deselect" the *y*-axis title (see Fig. 6.18).

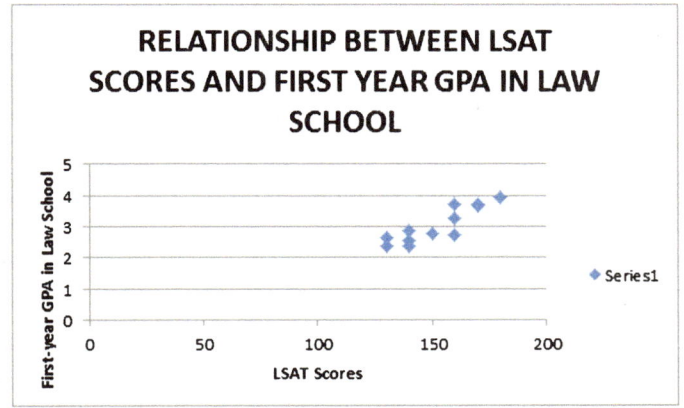

Fig. 6.18 Example of a chart title, an *x*-axis title, and a *y*-axis title

Legend (at top of screen)
None (to turn off the legend "series 1" at the far right side of the chart)
Gridlines (at top of screen)
Primary Horizontal Gridlines
None (to deselect the horizontal gridlines on the chart)

6.3.1.1   Moving the Chart Below the Table in the Spreadsheet

Objective: To move the chart below the table

Left-click your mouse on *any white space to the right of the top title inside the chart*, keep the left-click down and drag the chart down and to the left so that the top-left corner of the chart is in cell A26, then take your finger off the left-click of the mouse (see Fig. 6.19).

LAW SCHOOL ADMISSION TEST (LSAT)

Is there a relationship between LSAT scores and first-year GPA in law school?

LSAT score	First-year Law School GPA
130	2.65
170	3.72
140	2.85
160	3.25
150	2.75
180	3.95
130	2.35
160	2.74
170	3.65
140	2.55
160	3.72
140	2.35

n	12	12
mean	152.50	3.04
stdev	16.58	0.58
correlation		0.89

Fig. 6.19 Example of moving the chart below the table

6.3.1.2   Making the Chart "Longer" so that It Is "Taller"

Objective: To make the chart "longer" so that it is taller

Left-click your mouse on the bottom center of the chart to create an "up-and-down" arrow" sign, hold the left-click of the mouse down and drag the bottom of the chart down to row 48 to make the chart longer and then take your finger off the mouse.

6.3.1.3   Making the Chart "Wider"

> Objective: To make the chart "wider"

Put the pointer at the middle of the right border of the chart to create a "left-to-right arrow" sign and then left-click your mouse and hold the left-click down while you drag the right border of the chart to the middle of column H to make the chart wider (see Fig. 6.20).

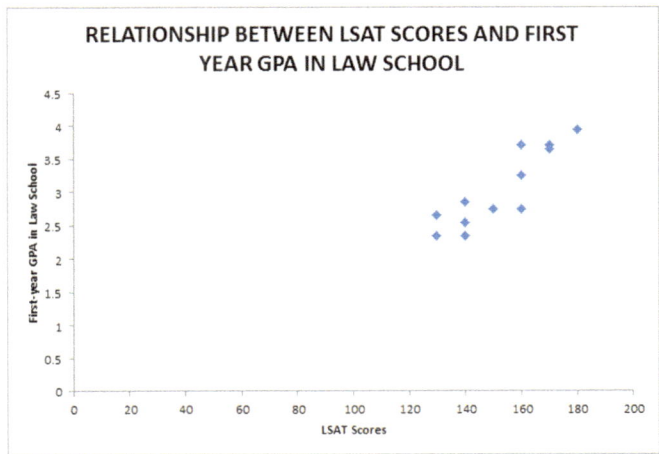

Fig. 6.20   Example of a chart that is enlarged to fit the cells A26:H48

Save this file as: LSAT5.

Note: If you printed LSAT5 now, it would "dribble over" to four pages of printout because the scale needs to be reduced below 100% in order for this spreadsheet to fit onto only one page.

Now, let us draw the regression line onto the chart. This regression line is called the "least-squares regression line," and it is the "best-fitting" straight line through the data points.

6.3.1.4   Drawing the Regression Line Through the Data Points in the Chart

> Objective: To draw the regression line through the data points on the chart

Right-click on any one of the data points inside the chart.

Add Trendline (see Fig. 6.21).

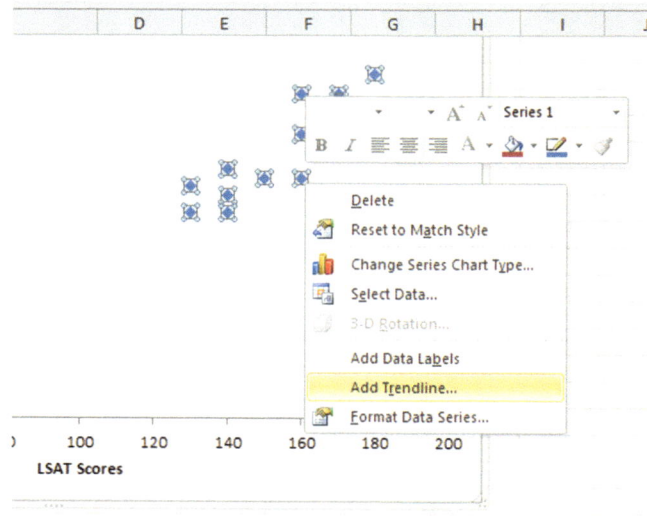

Fig. 6.21 Dialog box for adding a Trendline to the chart

Linear (be sure the "linear" button on the left is selected; see Fig. 6.22)

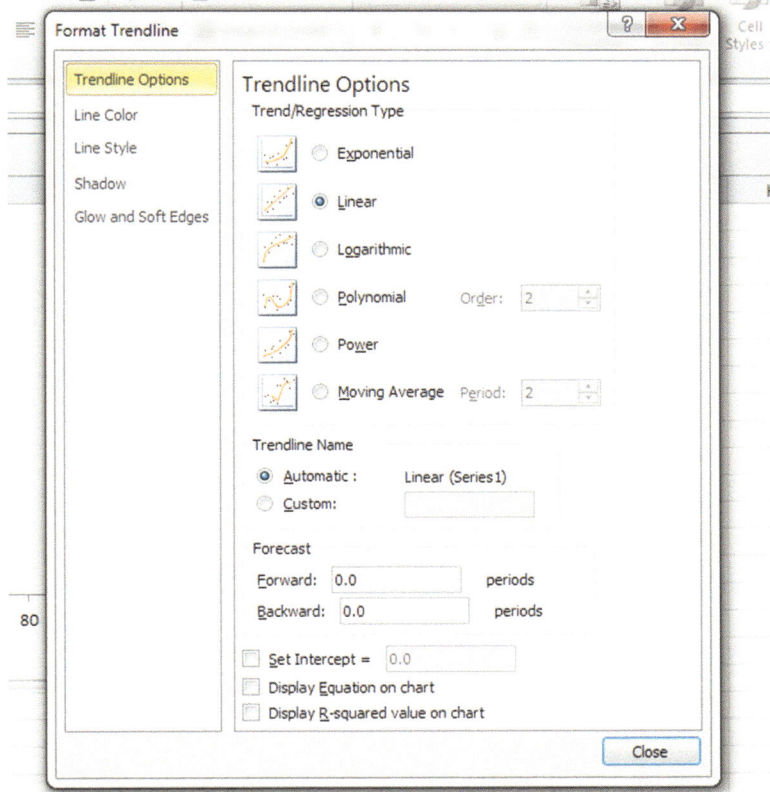

Fig. 6.22 Dialog box for a Linear Trendline

Close

Now, click on any blank cell outside the chart to "deselect" the chart.

Save this file as: LSAT6.

Note: If you printed this spreadsheet now, it is "too big" to fit onto one page and would "dribble over" onto four pages of printout because the scale needs to be reduced below 100% in order for this worksheet to fit onto only one page. You need to complete these next steps below to print out some, or all, of this spreadsheet.

6.4   Printing a Spreadsheet so that the Table and Chart Fit onto One Page

Objective: To print the spreadsheet so that the table and the chart fit onto one page

Page Layout (top of screen)

Change the scale at the middle icon near the top of the screen "Scale to Fit" by clicking on the down arrow until it reads "85%" so that the table and the chart will fit onto one page on your printout (see Fig. 6.23).

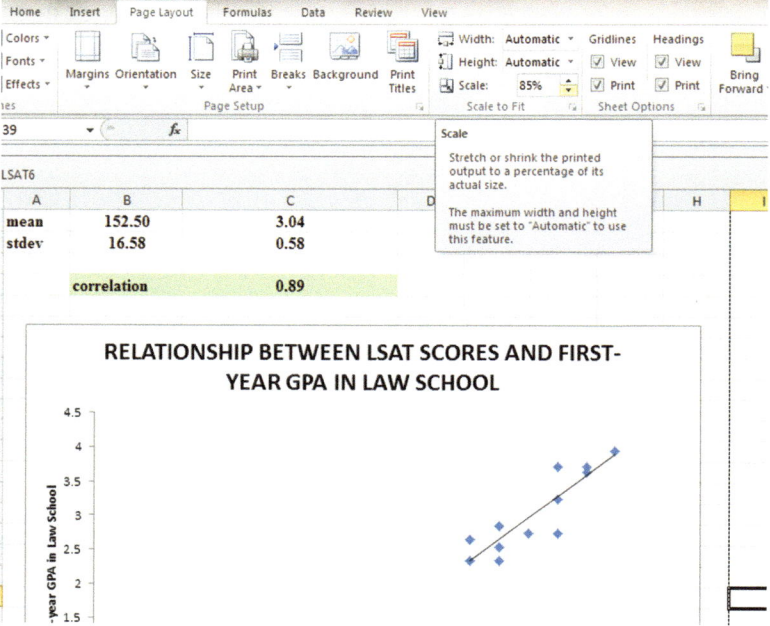

Fig. 6.23 Example of the Page Layout for reducing the scale of the chart to 85% of normal size

File
Print
Print (see Fig. 6.24)

LAW SCHOOL ADMISSION TEST (LSAT)

Is there a relationship between LSAT scores and first-year GPA in law school?

LSAT score	First-year Law School GPA
130	2.65
170	3.72
140	2.85
160	3.25
150	2.75
180	3.95
130	2.35
160	2.74
170	3.65
140	2.55
160	3.72
140	2.35

n	12	12
mean	152.50	3.04
stdev	16.58	0.58
correlation		0.89

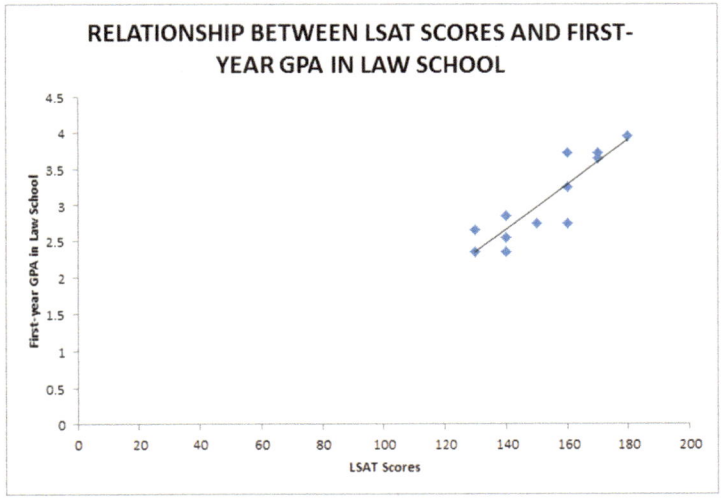

Fig. 6.24 Final spreadsheet of regression line on a chart (85% scale-to-fit size)

Resave your file as LSAT7.

6.5   Finding the Regression Equation

The main reason for charting the relationship between *X* and *Y* (i.e., LSAT scores as *X* and first-year GPA in Law School as *Y* in our example) is to see if there is a strong enough relationship between *X* and *Y* so that the regression equation that summarizes this relationship can be used to predict *Y* for a given value of *X*.

Since we know that the correlation between LSAT scores and GPA is +.89, this tells us that it makes sense to use LSAT scores to predict first-year GPA in Law School based on past data from this Law School.

We now need to find that regression equation that is the equation of the "best-fitting straight line" through the data points.

> Objective: To find the regression equation summarizing the relationship between *X* and *Y*.

In order to find this equation, we need to check to see if your version of Excel contains the "Data Analysis ToolPak" necessary to run a regression analysis.

6.5.1   *Installing the Data Analysis ToolPak into Excel*

> Objective: To install the Data Analysis ToolPak into Excel

Since there are currently three versions of Excel in the marketplace (2003, 2007, and 2010), we will give a brief explanation of how to install the Data Analysis ToolPak into each of these versions of Excel.

6.5.1.1   Installing the Data Analysis ToolPak into Excel 2010

Open a new Excel spreadsheet.
Click on: Data (at the top of your screen).

Look at the top of your monitor screen. Do you see the words "Data Analysis" at the far right of the screen? If you do, the Data Analysis ToolPak for Excel 2010 was correctly installed when you installed Office 2010, and you should skip ahead to Sect. 6.5.2.

If the words "Data Analysis" are not at the top right of your monitor screen, then the ToolPak component of Excel 2010 was not installed when you installed Office 2010 onto your computer. If this happens, you need to follow these steps:

File
Options

Excel options (creates a dialog box)

Add-Ins

Manage Excel Add-Ins (at the bottom of the dialog box)

Go

Highlight Analysis ToolPak (in the Add-ins dialog box)

OK

Data

(You now should have the words "Data Analysis" at the top right of your screen).

If you get a prompt asking you for the "installation CD," put this CD in the CD drive
and click on: OK.

Note: If these steps do not work, you should try these steps instead: File/Options
(bottom left)/Add-Ins/Analysis ToolPak/Go/click to the left of Analysis ToolPak
to add a check mark/OK.

If you need help doing this, ask your favorite "computer techie" for help.

You are now ready to skip ahead to Sect. 6.5.2.

6.5.1.2   Installing the Data Analysis ToolPak into Excel 2007

Open a new Excel spreadsheet.

Click on: Data (at the top of your screen).

If the words "Data Analysis" do not appear at the top right of your screen, you need
to install the Data Analysis ToolPak using the following steps:

Microsoft Office button (top left of your screen)

Excel options (bottom of dialog box)

Add-Ins (far left of dialog box)

Go (to create a dialog box for Add-Ins)

Highlight: Analysis ToolPak

OK (If Excel asks you for permission to proceed, click on Yes).

Data (You should now have the words "Data Analysis" at the top right of your
screen).

If you need help doing this, ask your favorite "computer techie" for help.

You are now ready to skip ahead to Sect. 6.5.2.

6.5.1.3   Installing the Data Analysis ToolPak into Excel 2003

Open a new Excel spreadsheet.

Click on: Tools (at the top of your screen).

If the bottom of this Tools box says "Data Analysis," the ToolPak has already been
installed in your version of Excel, and you are ready to find the regression
equation. If the bottom of the Tools box does not say "Data Analysis," you
need to install the ToolPak as follows:

Click on: File

Options (bottom left of screen)

Add-Ins

Analysis ToolPak (It is directly underneath Inactive Application Add-Ins near the top of the box).

Go

Click to add a check mark to the left of Analysis ToolPak

OK

Note: If these steps do not work, try these steps instead: Tools/Add-Ins/Click to the left of Analysis ToolPak to add a check mark to the left/OK.

You are now ready to skip ahead to Sect. 6.5.2.

6.5.2   Using Excel to Find the SUMMARY OUTPUT of Regression

You have now installed *ToolPak*, and you are ready to find the regression equation for the "best-fitting straight line" through the data points by using the following steps:

Open the Excel file *LSAT7* (if it is not already open on your screen).

Note: *If this file is already open, and there is a gray border around the chart, you need to click on any empty cell outside of the chart to deselect the chart.*

Now that you have installed *ToolPak*, you are ready to find the regression equation summarizing the relationship between LSAT scores and first-year GPA in Law School in your data set.

Remember that you gave the name *LSAT* to the *X* data (the predictor) and the name *GPA* to the *Y* data (the criterion) in a previous section of this chapter (see Sect. 6.2).

Data (top of screen)

Data Analysis (far right at top of screen; see Fig. 6.25)

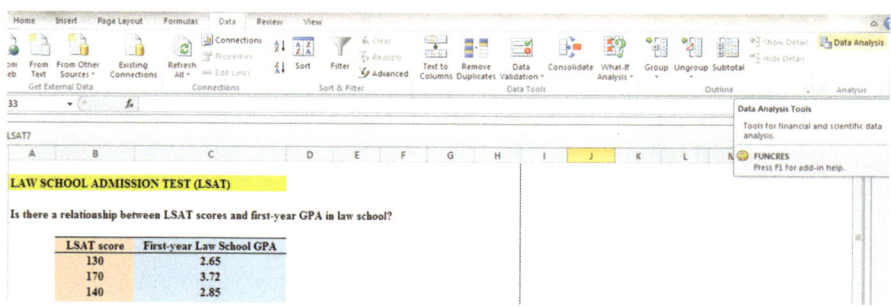

Fig. 6.25 Example of using the Data/Data Analysis function of Excel

Scroll down the dialog box using the down arrow and click on Regression
   (see Fig. 6.26).

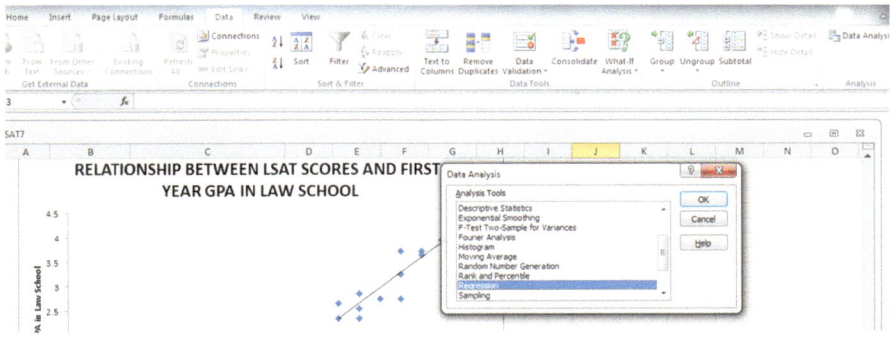

Fig. 6.26 Dialog box for creating the Regression function in Excel

OK
Input *Y* Range: GPA.
Input *X* Range: LSAT.
Click on the "button" to the left of Output Range to select this, and enter A50 in the
   box as the place on your spreadsheet to insert the Regression analysis in cell
   A50.
OK.
The *SUMMARY OUTPUT* should now be in cells A50: I67.
Now, make the columns in the Regression Summary Output section of your
   spreadsheet *wider* so that you can read all of the column headings clearly.
Now, change the data in the following three cells to number format (two decimal
   places):

B53
B66
B67

Now, change the format for all other numbers that are in decimal format to number
   format, three decimal places, and center all numbers within their cells.
Save the resulting file as: LSAT8.
Print the file so that it fits onto one page. (*Hint: Change the scale under "Page
   Layout" to 65% to make it fit.*) Your file should be like the file in Fig. 6.27.

Is there a relationship between LSAT scores and first-year GPA in law school?

LSAT score	First-year Law School GPA
130	2.65
170	3.72
140	2.85
160	3.25
150	2.75
180	3.95
130	2.35
160	2.74
170	3.65
140	2.55
160	3.72
140	2.35

n	12	12
mean	152.50	3.04
stdev	16.58	0.58
correlation		0.89

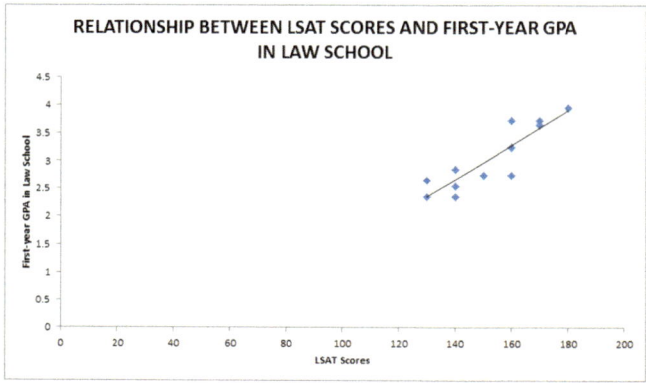

RELATIONSHIP BETWEEN LSAT SCORES AND FIRST-YEAR GPA IN LAW SCHOOL

SUMMARY OUTPUT

Regression Statistics	
Multiple R	0.89
R Square	0.787
Adjusted R Square	0.766
Standard Error	0.281
Observations	12

ANOVA

	df	SS	MS	F	Significance F
Regression	1	2.932	2.932	37.037	0.000
Residual	10	0.792	0.079		
Total	11	3.723			

	Coefficients	Standard Error	t Stat	P-value	Lower 95%	Upper 95%	Lower 95.0%	Upper 95.0%
Intercept	-1.70	0.784	-2.172	0.055	-3.451	0.044	-3.451	0.044
X Variable 1	0.03	0.005	6.086	0.000	0.020	0.043	0.020	0.043

Fig. 6.27 Final spreadsheet of correlation and simple linear regression including the summary output for the data

Note the following problem with the summary output.

Whoever wrote the computer program for this version of Excel made a mistake and gave the name "Multiple R" to cell A53. This is not correct. Instead, cell A53 should say "correlation r" since this is the notation that we are using for the correlation between X and Y.

You can now use your printout of the regression analysis to find the regression equation that is the best-fitting straight line through the data points.

But first, let us review some basic terms.

6.5.2.1 Finding the *y*-Intercept, *a*, of the Regression Line

The point on the *y*-axis that the regression line would intersect the *y*-axis if it were extended to reach the *y*-axis is called the "*y*-intercept," and *we will use the letter "a" to stand for the y-intercept of the regression line*. The *y*-intercept on the SUMMARY OUTPUT on the previous page is *−1.70 and appears in cell B66* (note the minus sign). This means that if you were to draw an imaginary line continuing down the regression line toward the *y*-axis that this imaginary line would cross the *y*-axis at −1.70. This is why *a* is called the "*y*-intercept."

6.5.2.2 Finding the Slope, *b*, of the Regression Line

The "tilt" of the regression line is called the "slope" of the regression line. It summarizes to what degree the regression line is either above or below a horizontal line through the data points. If the correlation between *X* and *Y* were zero, the regression line would be exactly horizontal to the *X*-axis and would have a zero slope.

If the correlation between X and Y is positive, the regression line would "slope upward to the right" above the X-axis. Since the regression line in Fig. 6.27 slopes upward to the right, the slope of the regression line is *+0.03 as given in cell B67. We will use the notation "b" to stand for the slope of the regression line*. (Note that Excel calls the slope of the line "*X* Variable 1" in the Excel printout).

Since the correlation between the LSAT scores and first-year GPA was *+.89*, you can see that the regression line for these data "slopes upward to the right" through the data. Note that the SUMMARY OUTPUT of the regression line in Fig. 6.27 gives a correlation, *r*, of +.89 in cell *B53*.

If the correlation between X and Y were negative, the regression line would "slope down to the right" above the X-axis. This would happen whenever the correlation between *X* and *Y* is a negative correlation that is between zero and minus one (0 and −1).

6.5.3 Finding the Equation for the Regression Line

To find the regression equation for the straight line that can be used to predict first-year GPA in Law School from an LSAT score, we only need two numbers in the SUMMARY OUTPUT in Fig. 6.27, *B66 and B67*.

The format for the regression line is

$$Y = a + b X \qquad (6.3)$$

where $a =$ *the y-intercept* (-1.70 in our example in cell B66) and $b =$ *the slope of the line* (+0.03 in our example in cell B67).

Therefore, the equation for the best-fitting regression line for our example is

$$Y = a + bX$$

$$\boxed{Y = -1.70 + 0.03X}.$$

Remember that Y is the first-year GPA that we are trying to predict, using the LSAT scores as the predictor, X.

Let us try an example using this formula to predict first-year GPA for a hypothetical student.

6.5.4   Using the Regression Line to Predict the Y-Value for a Given X-Value

Objective: To find the first-year GPA predicted from an LSAT score of 150

Since the LSAT score is 150 (i.e., $X = 150$), substituting this number into our regression equation gives:

$$Y = -1.70 + 0.03(150)$$

$$Y = -1.70 + 4.5$$

$$Y = 2.80$$

Important note: *If you look at your chart, if you go directly upward for an LSAT score of 150 until you hit the regression line, you see that you hit this line just under the number 3 on the y-axis to the left when you draw a line horizontal to the x-axis (actually, it is 2.80), the result above for predicting first-year GPA from an LSAT score of 150.*

Now, let us do a second example and predict what the first-year GPA would be if we used an LSAT score of 170.

$$Y = -1.70 + 0.03 X$$

$$Y = -1.70 + 0.03(170)$$

$$Y = -1.70 + 5.1$$

$$Y = 3.40$$

Important note: *If you look at your chart, if you go directly upward from an LSAT score of 170 until you hit the regression line, you see that you hit this line just under the number 3.5 on the y-axis to the left (actually it is 3.40), the result above for predicting first-year GPA from this LSAT score of 170.*

For a more detailed discussion of regression, see Black (2010).

6.6 Adding the Regression Equation to the Chart

Objective: To Add the Regression Equation to the Chart

If you want to include the regression equation within the chart next to the regression line, you can do that, but a word of caution first.

Throughout this book, we are using the regression equation for one predictor and one criterion to be the following:

$$Y = a + b X \tag{6.3}$$

where a = y-intercept and b = slope of the line.

See, for example, the regression equation in Sect. 6.5.3 where the y-intercept was $a = -1.70$, and the slope of the line was $b = +0.03$ to generate the following regression equation:

$$Y = -1.70 + 0.03 X$$

However, Excel 2010 uses a slightly different regression equation (which is logically identical to the one used in this book) when you add a regression equation to a chart:

$$Y = b X + a \tag{6.4}$$

where a = y-intercept and b = slope of the line.

Note that this equation is identical to the one we are using in this book with the terms arranged in a different sequence.

For the example we used in Sect. 6.5.3, Excel 2010 would write the regression equation on the chart as:

$$Y = 0.03 X - 1.70$$

This is the format that will result when you add the regression equation to the chart using Excel 2010 using the following steps:

Open the file LSAT8 (that you saved in Sect. 6.5.2).

Click just *inside* the outer border of the chart in the top right corner to add the "gray border" around the chart in order to "select the chart" for changes you are about to make.

Right-click on any of the data points in the chart.

Highlight: Add Trendline.

The "Linear button" near the top of the dialog box will be selected (on its left).

Click on: Display Equation on chart (near the bottom of the dialog box; see Fig. 6.28).

Close

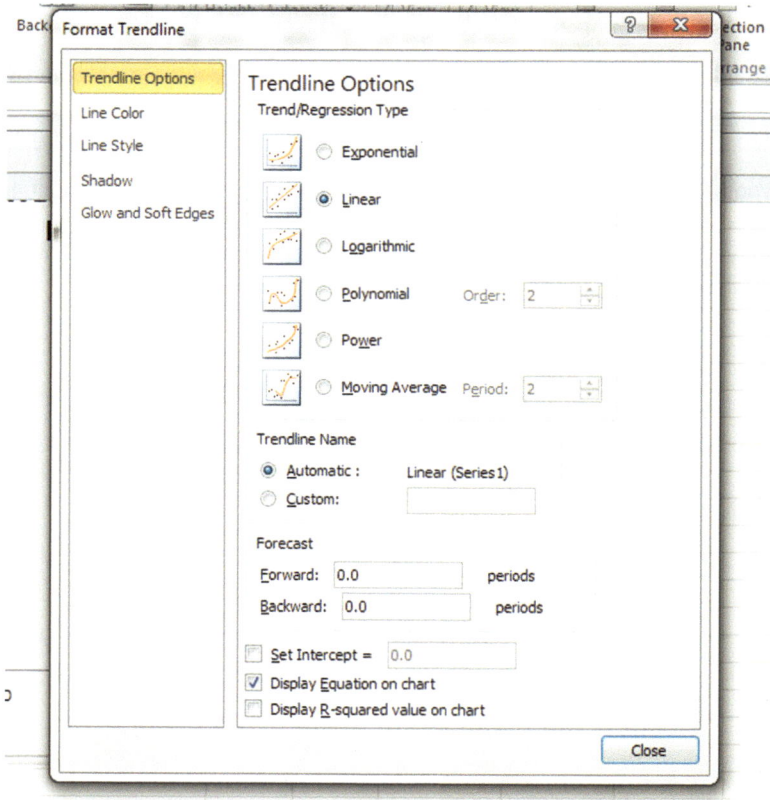

Fig. 6.28 Dialog box for adding the regression equation to the chart next to the regression line on the chart

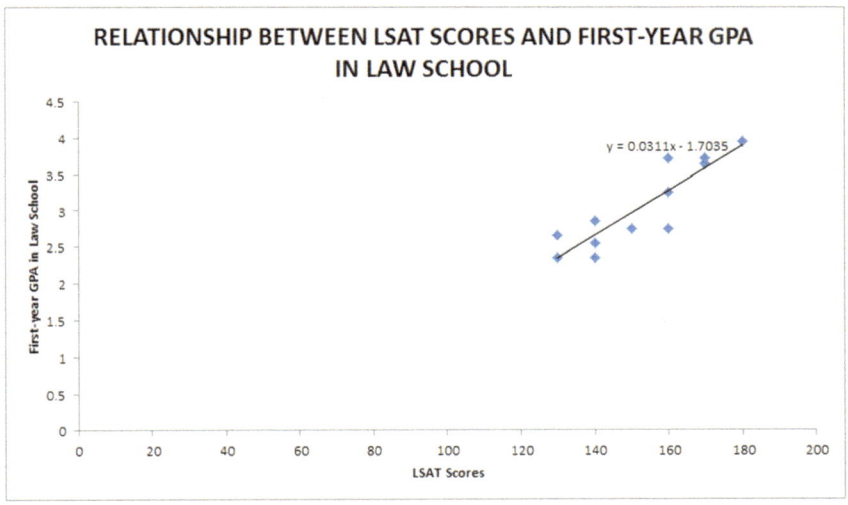

Fig. 6.29 Example of a chart with the regression equation displayed next to the regression line

Note that the regression equation on the chart is in the following form next to the
regression line on the chart (see Fig. 6.29):

$$Y = 0.03 X - 1.70$$

Now, save this file as: LSAT9.

6.7   How to Recognize Negative Correlations
in the SUMMARY OUTPUT Table

Important note: *Since Excel does not recognize negative correlations in the
SUMMARY OUTPUT results, but treats all correlations as if
they were positive correlations (this was a mistake made by the
programmer), you need to be careful to note that there may be a
negative correlation between X and Y even if the printout says
that the correlation is a positive correlation.*

*You will know that the correlation between X and Y is a
negative correlation when these two things occur:*

1. *The slope, b, is a negative number. This can only occur when there is a negative
correlation.*
2. *The chart clearly shows a downward slope in the regression line, which can only
occur when the correlation between X and Y is negative.*

6.8 Printing Only Part of a Spreadsheet Instead of the Entire Spreadsheet

> Objective: To print part of a spreadsheet separately instead of printing the entire spreadsheet

There will be many occasions when your spreadsheet is so large in the number of cells used for your data and charts that you only want to print part of the spreadsheet separately so that the print will not be so small that you cannot read it easily.

We will now explain how to print only part of a spreadsheet onto a separate page by using three examples of how to do that using the file, LSAT9, that you created in Sect. 6.6: (1) printing only the table and the chart on a separate page, (2) printing only the chart on a separate page, and (3) printing only the SUMMARY OUTPUT of the regression analysis on a separate page.

Note: If the file LSAT9 is not open on your screen, you need to open it now.

Let us describe how to do these three goals with three separate objectives:

6.8.1 Printing Only the Table and the Chart on a Separate Page

> Objective: To print only the table and the chart on a separate page

1. Left-click your mouse starting at the top left of the table *in cell A2* and drag the mouse *down and to the right so that all of the table and all of the chart are highlighted in light blue on your computer screen from cell A2 to cell H48* (the light-blue cells are called the "selection" cells).
2. File
   Print
   Print Active Sheet (hit the down arrow on the right)
   Print selection
   Print

The resulting printout should contain only the table of the data and the chart resulting from the data.

Then, click on any empty cell in your spreadsheet to deselect the table and chart.

6.8.2   *Printing Only the Chart on a Separate Page*

Objective: To print only the chart on a separate page
---

1. Click on any "white space" *just inside the outside border of the chart in the top right corner of the chart* to create the gray border around all of the borders of the chart in order to "select" the chart.
2. File
   Print
   Print selected chart
   Print selected chart (again)
   Print

The resulting printout should contain only the chart resulting from the data.

Important note: *After each time you print a chart by itself on a separate page, you should immediately click on any white space OUTSIDE the chart to remove the gray border from the border of the chart. When the gray border is on the borders of the chart, this tells Excel that you want to print only the chart by itself. You should do this now!*

6.8.3   *Printing Only the SUMMARY OUTPUT of the Regression Analysis on a Separate Page*

Objective: To print only the SUMMARY OUTPUT of the regression analysis on a separate page
---

1. Left-click your mouse at the cell just above SUMMARY OUTPUT in *cell A49* on the left of your spreadsheet and drag the mouse *down and to the right* until all of the regression output is highlighted in dark blue on your screen from A49 to I67.
2. File
   Print
   Print active sheets (hit the down arrow on the right)
   Print selection
   Print

The resulting printout should contain only the summary output of the regression analysis on a separate page.

Finally, click on any empty cell on the spreadsheet to "deselect" the regression table.

6.9 End-of-Chapter Practice Problems

1. Suppose that you have been asked to analyze some data from the SAT Reasoning Test (formerly called the Scholastic Aptitude Test), which is a standardized test for college admissions in the USA. This test is intended to measure a student's readiness for academic work in college, and about 1.4 million high school students take this test every year. There are three subtest scores generated from this test: critical reading, writing, and mathematics, and each of these subtests has a score range between 200 and 800 with an average score of about 500.

   Suppose that a selective liberal arts college in the northeast of the USA wants to determine the relationship between the SAT Reading score and freshman grade-point average (GPA) at the end of freshman year at this college, and that this college has asked you to determine this relationship.

   You have decided to use the SAT Reading score as the predictor, *X*, and the freshman grade-point average (GPA) as the criterion, *Y*. To test your Excel skills, you have selected ten students randomly from last year's freshmen class and have recorded their scores on these variables in Fig. 6.30.

SAT REASONING TEST (SAT)

How well can the SAT Critical Reading score predict FROSH GPA?

SAT Critical Reading score	FROSH GPA
250	2.23
400	2.15
600	3.65
500	2.64
520	2.45
720	3.67
630	3.46
550	2.86
650	3.55
630	3.15

Fig. 6.30 Worksheet data for Chap. 6: practice problem #1

Create an Excel spreadsheet and enter the data *using the reading score as the independent variable (predictor) and FROSH GPA as the dependent variable (criterion).*

Important note: *When you are trying to find a correlation between two variables, it is important that you place the predictor, X, ON THE LEFT COLUMN in your Excel spreadsheet, and the criterion, Y,*

IMMEDIATELY TO THE RIGHT OF THE X COLUMN. You should do this every time that you want to use Excel to find a correlation between two variables to check your thinking so that you do not confuse these two variables with one another.

(a) Use Excel's =correl function to find the correlation between these two variables and round off the result to two decimal places.

(b) Create an *XY scatterplot* of these two sets of data such that:

- Top title: RELATIONSHIP BETWEEN SAT READING SCORE AND FROSH GPA
- *x*-Axis title: SAT Critical Reading Score
- *y*-Axis title: FROSH GPA
- Resize the chart so that it is 8-columns wide and 25-rows long
- Delete the legend
- Delete the gridlines
- Move the chart below the table

(c) Create the *least-squares regression line* for these data on the scatterplot and add the regression equation to the chart.

(d) Use Excel to run the regression statistics to find the *equation for the least-squares regression line* for these data and display the results below the chart on your spreadsheet. Use number format (two decimal places) for the correlation and three decimal places for all the other decimal numbers.

(e) Print just the input data and the chart so that this information fits onto one page. Then, print the regression output table on a separate page so that it fits onto that separate page.

(f) Save the file as SAT4.

Now, answer these questions using your Excel printout:

1. What is the *y*-intercept?
2. What is the slope of the line?
3. What is the regression equation for these data (use three decimal places for the *y*-intercept and the slope)?
4. Use the regression equation to predict the FROSH GPA you would expect for a SAT Reading Score of 600.

2. Suppose that you have been hired as a consultant to determine if there is a relationship between the on-time performance of major airlines and the number of passenger complaints lodged by passengers in the US Department of Transportation. These data can be found in the Air Travel Consumer Report published by the Office of Aviation Enforcement and Proceedings of the US Department of Transportation and in *The Wall Street Journal* (McCartney 2010; these data are presented in Fig. 6.31). Note that the data for passenger complaints are converted

How the Major Airlines Performed in 2009

Airline	% On-time Arrivals	Passenger complaints per million passengers
Southwest	82.5	2.1
Alaska	82.4	5.5
United	80.5	13.4
US Airways	80.5	13.1
Delta	78.6	16.7
Continental	77.9	10.1
JetBlue	77.2	8.9
AirTran	76.0	9.9
American	75.7	10.8

Fig. 6.31 Worksheet data for Chap. 6: practice problem #2

to a scale "per million passengers" so that all of the airlines can be measured on the same scale, regardless of the number of passengers they flew.

Create an Excel spreadsheet and enter the data using % On-time Arrivals as the independent variable (predictor) and Passenger Complaints per million passengers as the dependent variable (criterion). Be sure to enter the data for on-time percent *as numbers, and not as decimals.* For example, an on-time percent of 82.5 should be entered on your spreadsheet as 82.5 and not as 0.825.

(a) Create an *XY scatterplot* of these two sets of data such that:

- Top title: RELATIONSHIP BETWEEN ON-TIME% AND PASSEN-GER COMPLAINTS
- *x*-Axis title: % On-time Arrivals
- *y*-Axis title: Passenger Complaints per million passengers
- Resize the chart so that it is 7-columns wide and 25-rows long
- Delete the legend
- Delete the gridlines
- Move the chart below the table

(b) Create the *least-squares regression line* for these data on the scatterplot.
(c) Use Excel to run the regression statistics to find the *equation for the least-squares regression line* for these data and display the results below the chart on your spreadsheet. Use number format (two decimal places) for the correlation, r, and for both the y-intercept and the slope of the line.
(d) Print the input data and the chart so that this information fits onto one page.
(e) Then, print out the regression output table so that this information fits onto a separate page.

By hand:

(1a) Circle and label the value of the *y-intercept* and the *slope* of the regression line onto that separate page.

(2b) *Read from the graph* the number of passenger complaints you would predict for an *on-time arrival rate of 77%* and write your answer in the space immediately below:

_____

(f) Save the file as ontime6.
Answer the following questions using your Excel printout:

1. What is the correlation?
2. What is the *y*-intercept?
3. What is the slope of the line?
4. What is the regression equation for these data (use two decimal places for the *y*-intercept and the slope)?
5. Use that regression equation to predict the passenger complaints you would expect for an airline with an on-time arrival of 80%.

(Note that this correlation is not the multiple correlation as the Excel table indicates, but it is merely the correlation r instead).

Important note: *Since Excel does not recognize negative correlations in the SUMMARY OUTPUT but treats all correlations as if they were positive correlations, you need to be careful to note that there is a negative correlation between on-time performance and passenger complaints.*

You know this for two reasons:

1. *The slope, b, is a negative −0.69 which can only occur when there is a negative correlation.*
2. *The chart clearly shows a downward slope in the regression line, which can only happen when the correlation is negative.*

Therefore, the correlation between on-time percent and complaints per million passengers is not +0.41, but −0.41 for this problem. This is a negative correlation!

3. Suppose that a tenth grade Geometry teacher asks you to find the relationship between the number of school days her students were absent during the year and the scores of her students on an end-of-year standardized test of Geometry. Before you do this analysis, you want to check your understanding of Excel on a small group of students, and so you collect the data on 11 students selected randomly from her class. These hypothetical data appear in Fig. 6.32.

End-of-year Math test results

Is there a relationship between absentee rate and Math Test scores?

No. days absent this year	Math Test score
6	92
7	88
21	65
10	75
8	72
9	83
10	86
14	73
16	65
12	71
17	67

Fig. 6.32 Worksheet data for Chap. 6: practice problem #3

Create an Excel spreadsheet and enter the data using the number of days absent as the independent variable (predictor) and the Math Test Score as the dependent variable (criterion).

(a) Create an *XY scatterplot* of these two sets of data such that:

- Top title: RELATIONSHIP BETWEEN DAYS ABSENT AND MATH TEST SCORE
- *x*-Axis title: No. days absent
- *y*-Axis title: Math Test Score
- Move the chart below the table
- Resize the chart so that it is 8 columns wide and 25 rows long
- Delete the legend
- Delete the gridlines

(b) Create the *least-squares regression line* for these data on the scatterplot and add the regression equation to the chart.

(c) Use Excel to run the regression statistics to find the *equation for the least-squares regression line* for these data and display the results below the chart on your spreadsheet. Use number format (two decimal places) for the correlation and the two coefficients and use three decimal places for all other decimal figures.

(d) Print just the input data and the chart so that this information fits onto one page. Then, print the regression output table on a separate page so that it fits onto that separate page.

(e) Save the file as ABSENT6.

Answer the following questions using your Excel printout:

1. What is the correlation between the number of days absent and Geometry Test Scores?
2. What is the y-intercept?
3. What is the slope of the line?
4. What is the regression equation?
5. Use the regression equation to predict the Geometry Test Score you would expect for a student who had been absent for 15 days during the school year. Show your work on a separate sheet of paper.

References

Black, K. Business Statistics: For Contemporary Decision Making (6th ed.). Hoboken, NJ: John Wiley & Sons, Inc., 2010.

Levine, D.M.. Stephan, D.F., Krehbiel, T.C., and Berenson, M.L. Statistics for Managers Using Microsoft Excel (6th ed.). Boston, MA: Prentice Hall/Pearson, 2011.

McCartney, S. An airline report card: fewer delays, hassles last year, but bumpy times may be ahead. *The Wall Street Journal* (2010 January 7), pp. D1, D3.

Zikmund, W.G. and Babin, B.J. Exploring Marketing Research (10th ed.). Mason, OH: South-Western Cengage Learning, 2010.

Chapter 7
Multiple Correlation and Multiple Regression

There are many times in business when you want to predict a criterion, Y, but you want to find out if you can develop a better prediction model by using *several predictors* in combination (e.g., X_1, X_2, X_3, etc.) instead of a single predictor, X.

The resulting statistical procedure is called "multiple correlation" because it uses two or more predictors in combination to predict Y, instead of a single predictor, X. Each predictor is "weighted" differently based on its separate correlation with Y and its correlation with the other predictors. The job of multiple correlation is to produce a regression equation that will weight each predictor differently and in such a way that the combination of predictors does a better job of predicting Y than any single predictor by itself. We will call the multiple correlation: R_{xy}.

You will recall (see Sect. 6.5.3) that the regression equation that predicts Y when only one predictor, X, is used is:

$$Y = a + bX \tag{7.1}$$

7.1 Multiple Regression Equation

The multiple regression equation follows a similar format and is:

$$Y = a + b_1X_1 + b_2X_2 + b_3X_3$$
$$+ etc. \ depending \ on \ the \ number \ of \ predictors \ used \tag{7.2}$$

The "weight" given to each predictor in the equation is represented by the letter "b" with a subscript to correspond to the same subscript on the predictors.

Important note: *In order to do multiple regression, you need to have installed the "Data Analysis ToolPak" that was described in Chap. 6 (see Sect. 6.5.1). If you did not install this, you need to do so now.*

T. Quirk, *Excel 2010 for Educational and Psychological Statistics:*
A Guide to Solving Practical Problems, DOI 10.1007/978-1-4614-2071-2_7,
© Springer Science+Business Media, LLC 2012

Let us try a practice problem.

Suppose that you have been asked to analyze some data from the SAT Reasoning Test (formerly called the Scholastic Aptitude Test) which is a standardized test for college admissions in the USA. This test is intended to measure a student's readiness for academic work in college, and about 1.4 million high school students take this test every year. There are three subtest scores generated from this test: Critical reading, writing, and mathematics, and each of these subtests has a score range between 200 and 800 with an average score of about 500.

Suppose that a nearby selective liberal arts college in the northeast of the USA that is near to you wants to determine the relationship between SAT reading scores, SAT writing scores, and SAT math scores in their ability to predict freshman grade-point average (FROSH GPA) at the end of freshman year at this college and that this college has asked you to determine this relationship.

You have decided to use the three subtest scores as the predictors, X_1, X_2, and X_3 and the freshman grade-point average (FROSH GPA) as the criterion, Y. To test your Excel skills, you have selected 11 students randomly from last year's freshmen class and have recorded their scores on these variables.

Let us use the following notation:

Y FROSH GPA
X_1 READING SCORE
X_2 WRITING SCORE
X_3 MATH SCORE

Suppose, further, that you have collected the following hypothetical data summarizing these scores (see Fig. 7.1).

	A	B	C	D	E
1					
2	SAT REASONING TEST				
3					
4	Is there a relationahip between SAT scores and Freshman GPA at a local college?				
5					
6	FROSH GPA	READING SCORE	WRITING SCORE	MATH SCORE	
7	2.55	250	230	220	
8	3.05	610	240	440	
9	3.55	620	540	530	
10	2.05	420	420	260	
11	2.45	320	520	320	
12	2.95	630	620	620	
13	3.15	650	540	530	
14	3.45	520	580	560	
15	3.30	420	490	630	
16	2.75	330	220	610	
17	3.65	440	570	660	
18					
19					

Fig. 7.1 Worksheet data for SAT vs. FROSH GPA (practical example)

Create an Excel spreadsheet for these data using the following cell reference:

A2:   SAT REASONING TEST
A4:   Is there a relationship between SAT scores and freshman GPA
            at a local college?
A6:   FROSH GPA
A7:   2.55
B6:   READING SCORE
C6:   WRITING SCORE
D6:   MATH SCORE
D17:   660

Next, change the column width to match the above table and change all GPA figures to number format (two decimal places).

Now, fill in the additional data in the chart such that

A17:   3.65
B17:   440
C17:   570

Then, center all numbers in your table.

Important note: *Be sure to double-check all of your numbers in your table to be sure that they are correct, or your spreadsheets will be incorrect.*

Save this file as: GPA15.

Before we do the multiple regression analysis, we need to try to make one important point very clear:

Important: *When we used one predictor, X, to predict one criterion, Y, we said that you need to make sure that the X variable is ON THE LEFT in your table, and the Y variable is ON THE RIGHT in your table so that you do not get these variables mixed up (see Sect. 6.3).*

However, in multiple regression, you need to follow this rule which is exactly the opposite:

When you use several predictors in multiple regression, it is essential that the criterion you are trying to predict, Y, be ON THE FAR LEFT, and all of the predictors are TO THE RIGHT of the criterion, Y, in your table so that you know which variable is the criterion, Y, and which variables are the predictors. If you make this a habit, you will save yourself a lot of grief.

Notice in the table above that the criterion Y (FROSH GPA) is on the far left of the table, and the three predictors (READING SCORE, WRITING SCORE, and MATH SCORE) are to the right of the criterion variable. If you follow this rule, you will be less likely to make a mistake in this type of analysis.

7.2   Finding the Multiple Correlation and the Multiple Regression Equation

Objective: To find the multiple correlation and multiple regression equation using Excel

You do this by the following commands:

Data.
Click on Data Analysis (far right top of screen).
Regression (scroll down to this in the box; see Fig. 7.2).

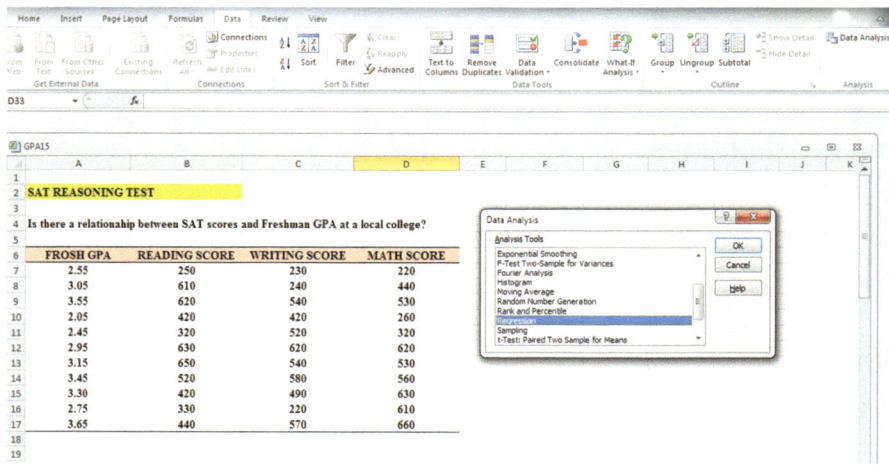

Fig. 7.2   Dialog box for regression function

OK.
Input *Y* range: A6:A17.
Input *X* range: B6:D17.
Click on the labels box to *add a check mark* to it (because you have included the column labels in row 6).
Output range (click on the button to its left, and enter): A20 (see Fig. 7.3).

Important note: *Excel automatically assigns a dollar sign $ in front of each column letter and each row number so that you can keep these ranges of data constant for the regression analysis.*

OK. (See Fig. 7.4 to see the resulting SUMMARY OUTPUT)

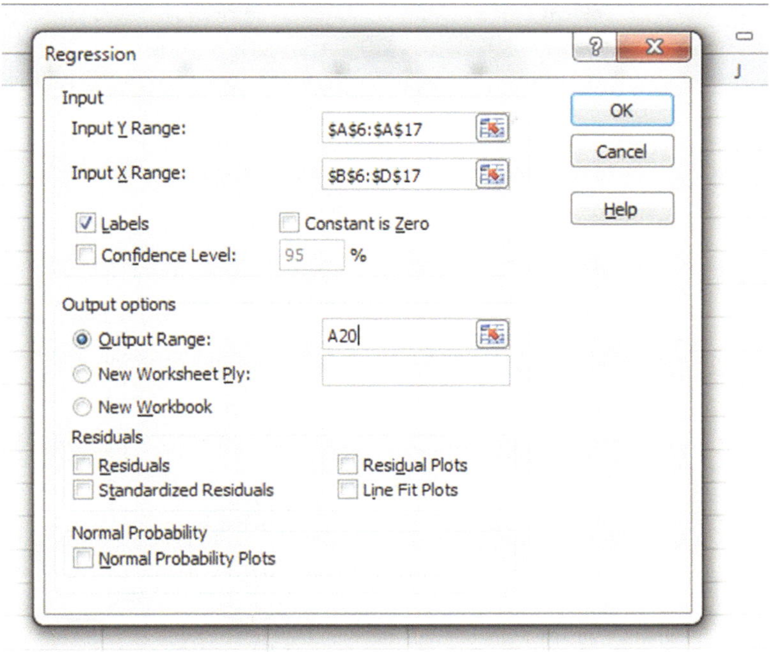

Fig. 7.3 Dialog box for SAT vs. FROSH GPA data

Next, format cell B23 in number format (two decimal places).

Next, format the following four cells in number format (four decimal places):

B36
B37
B38
B39

Change all other decimal figures to two decimal places, and center all figures within their cells.

Note that both the input *Y* range and the input *X* range above both include the label at the top of the columns.

Save the file as: GPA16.

Now, print the file so that it fits onto one page by changing the scale to *60% size*. The resulting regression analysis is given in Fig. 7.5.

Once you have the SUMMARY OUTPUT, you can determine the multiple correlation and the regression equation that is the best-fit line through the data points using READING SCORE, WRITING SCORE, AND MATH SCORE as the three predictors and FROSH GPA as the criterion.

Note on the SUMMARY OUTPUT where it says: "Multiple R." This term is correct since this is the term Excel uses for the multiple correlation, which is +0.80.

SAT REASONING TEST

Is there a relationship between SAT scores and Freshman GPA at a local college?

FROSH GPA	READING SCORE	WRITING SCORE	MATH SCORE
2.55	250	230	220
3.05	610	240	440
3.55	620	540	530
2.05	420	420	260
2.45	320	520	320
2.95	630	620	620
3.15	650	540	530
3.45	520	580	560
3.30	420	490	630
2.75	330	220	610
3.65	440	570	660

SUMMARY OUTPUT

Regression Statistics	
Multiple R	0.797651156
R Square	0.636247366
Adjusted R Square	0.48035338
Standard Error	0.361446932
Observations	11

ANOVA

	df	SS	MS	F	Significance F
Regression	3	1.599583719	0.533194573	4.081282	0.057174747
Residual	7	0.91450719	0.130643884		
Total	10	2.514090909			

	Coefficients	Standard Error	t Stat	P-value	Lower 95%	Upper 95%	Lower 95.0%	Upper 95.0%
Intercept	1.53627108	0.468442063	3.279532734	0.013496	0.428581617	2.643960543	0.428581617	2.643960543
READING SCORE	0.000642945	0.000963026	0.667629207	0.525762	-0.001634251	0.00292014	-0.001634251	0.00292014
WRITING SCORE	0.000264354	0.000889915	0.297055329	0.775046	-0.00183996	0.002368667	-0.00183996	0.002368667
MATH SCORE	0.00210733	0.000848684	2.4830572	0.042022	0.000100512	0.004114149	0.000100512	0.004114149

Fig. 7.4 Regression SUMMARY OUTPUT of SAT vs. FROSH GPA data

This means that from these data the combination of READING SCORES, WRITING SCORES, AND MATH SCORES together form a very strong positive relationship in predicting FROSH GPA.

To find the regression equation, *notice the coefficients at the bottom of the SUMMARY OUTPUT*:

Intercept: a (this is the y-intercept)	1.5363
READING SCORE: b_1	0.0006
WRITING SCORE: b_2	0.0003
MATH SCORE: b_3	0.0021

Since the general form of the multiple regression equation is:

$$Y = a + b_1 X_1 + b_2 X_2 + b_3 X_3 \qquad (7.2)$$

SAT REASONING TEST

Is there a relationahip between SAT scores and Freshman GPA at a local college?

FROSH GPA	READING SCORE	WRITING SCORE	MATH SCORE
2.55	250	230	220
3.05	610	240	440
3.55	620	540	530
2.05	420	420	260
2.45	320	520	320
2.95	630	620	620
3.15	650	540	530
3.45	520	580	560
3.30	420	490	630
2.75	330	220	610
3.65	440	570	660

SUMMARY OUTPUT

Regression Statistics	
Multiple R	0.80
R Square	0.64
Adjusted R Square	0.48
Standard Error	0.36
Observations	11

ANOVA

	df	SS	MS	F	Significance F
Regression	3	1.60	0.53	4.08	0.06
Residual	7	0.91	0.13		
Total	10	2.51			

	Coefficients	Standard Error	t Stat	P-value	Lower 95%	Upper 95%	Lower 95.0%	Upper 95.0%
Intercept	1.5363	0.47	3.28	0.01	0.43	2.64	0.43	2.64
READING SCORE	0.0006	0.00	0.67	0.53	0.00	0.00	0.00	0.00
WRITING SCORE	0.0003	0.00	0.30	0.78	0.00	0.00	0.00	0.00
MATH SCORE	0.0021	0.00	2.48	0.04	0.00	0.00	0.00	0.00

Fig. 7.5 Final spreadsheet for SAT vs. FROSH GPA regression analysis

we can now write the multiple regression equation for these data:
$$Y = 1.5363 + 0.0006X_1 + 0.0003X_2 + 0.0021X_3$$

7.3 Using the Regression Equation to Predict FROSH GPA

> Objective: To find the predicted FROSH GPA using an SAT Reading Score of 600, an SAT Writing Score of 500, and an SAT Math Score of 550

Plugging these three numbers into our regression equation gives us
$$Y = 1.5363 + 0.0006(600) + 0.0003(500) + 0.0021(550)$$
$$Y = 1.5363 + 0.36 + 0.15 + 1.155$$
$$Y = 3.20 \text{ (since GPA scores are typically measured in two decimals)}$$

If you want to learn more about the theory behind multiple regression, see Keller (2009).

7.4 Using Excel to Create a Correlation Matrix in Multiple Regression

The final step in multiple regression is to find the correlation between all of the variables that appear in the regression equation.

In our example, this means that we need to find the correlation between each of the six pairs of variables.

To do this, we need to use Excel to create a "correlation matrix." This matrix summarizes the correlations between all of the variables in the problem.

Objective: To use Excel to create a correlation matrix between the four variables in this example

---

To use Excel to do this, use these steps:

Data (top of screen under "Home" at the top left of screen).
Data analysis.
Correlation (scroll *up* to highlight this formula; see Fig. 7.6).
OK.
Input range: A6:D17.

(Note that this input range includes the labels at the top of the four variables (FROSH GPA, READING SCORE, WRITING SCORE, MATH SCORE) as well as all of the figures in the original data set).

Grouped by: columns.
Put a check in the box for: labels in the first row (since you included the labels at the top of the columns in your input range of data above).
Output range (click on the button to its left, and enter):   A42 (see Fig. 7.7).
OK.

The resulting correlation matrix appears in A42:E46 (see Fig. 7.8).

Next, format all of the numbers in the correlation matrix that are in decimals to two decimal places. And, also, make column E wider so that the MATH SCORE label fits inside cell E42.

Save this Excel file as: GPA14.

The final spreadsheet for these scores appears in Fig. 7.9.

Note that the number "1" along the diagonal of the correlation matrix means that the correlation of each variable with itself is a perfect, positive correlation of 1.0.

Correlation coefficients are always expressed in just two decimal places.

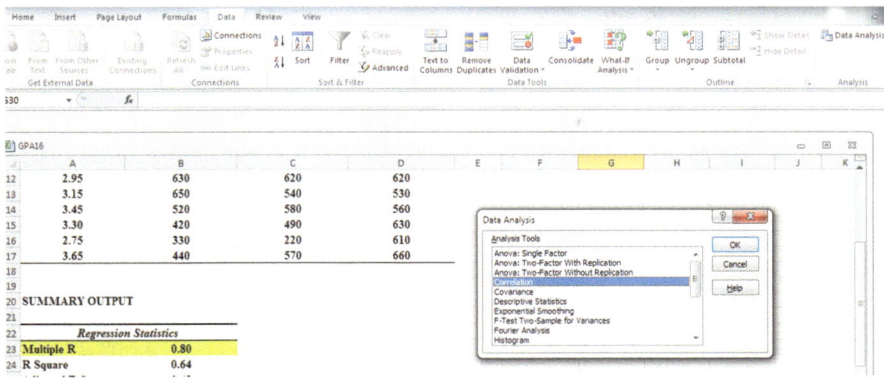

Fig. 7.6 Dialog box for SAT vs. FROSH GPA Correlations

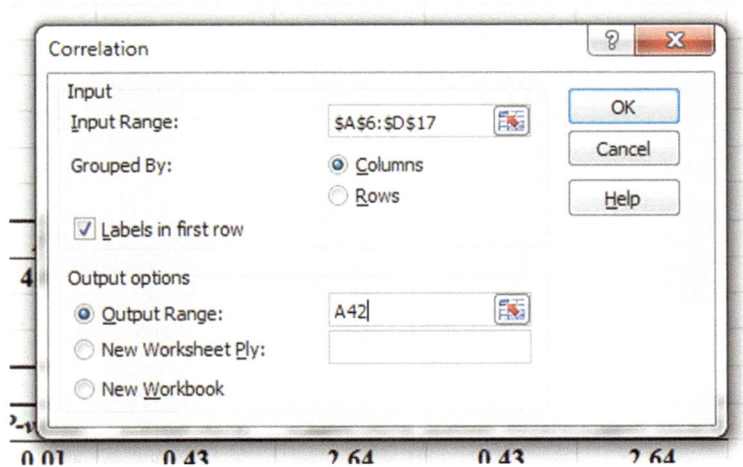

Fig. 7.7 Dialog box for Input/Output range for correlation matrix
OK

	FROSH GPA	READING SCORE	WRITING SCORE	1ATH SCORE
FROSH GPA	1			
READING SCORE	0.510369686	1		
WRITING SCORE	0.446857676	0.468105152	1	
MATH SCORE	0.772523347	0.444074496	0.429202393	1

Fig. 7.8 Resulting correlation matrix for SAT scores vs. FROSH GPA data

Is there a relationahip between SAT scores and Freshman GPA at a local college?

FROSH GPA	READING SCORE	WRITING SCORE	MATH SCORE
2.55	250	230	220
3.05	610	240	440
3.55	620	540	530
2.05	420	420	260
2.45	320	520	320
2.95	630	620	620
3.15	650	540	530
3.45	520	580	560
3.30	420	490	630
2.75	330	220	610
3.65	440	570	660

SUMMARY OUTPUT

Regression Statistics	
Multiple R	0.80
R Square	0.64
Adjusted R Square	0.48
Standard Error	0.36
Observations	11

ANOVA

	df	SS	MS	F	Significance F
Regression	3	1.60	0.53	4.08	0.06
Residual	7	0.91	0.13		
Total	10	2.51			

	Coefficients	Standard Error	t Stat	P-value	Lower 95%	Upper 95%	Lower 95.0%	Upper 95.0%
Intercept	1.5363	0.47	3.28	0.01	0.43	2.64	0.43	2.64
READING SCORE	0.0006	0.00	0.67	0.53	0.00	0.00	0.00	0.00
WRITING SCORE	0.0003	0.00	0.30	0.78	0.00	0.00	0.00	0.00
MATH SCORE	0.0021	0.00	2.48	0.04	0.00	0.00	0.00	0.00

	FROSH GPA	READING SCORE	WRITING SCORE	MATH SCORE
FROSH GPA	1			
READING SCORE	0.51	1		
WRITING SCORE	0.45	0.47	1	
MATH SCORE	0.77	0.44	0.43	1

Fig. 7.9 Final spreadsheet for SAT scores vs. FROSH GPA regression and the correlation matrix

You are now ready so read the correlation between the six pairs of variables:

The correlation between READING SCORE and FROSH GPA is:	+.51.
The correlation between WRITING SCORE and FROSH GPA is:	+.45.
The correlation between MATH SCORE and FROSH GPA is:	+.77.
The correlation between WRITING SCORE and READING SCORE is:	+.47.
The correlation between MATH SCORE and READING SCORE is:	+.44.
The correlation between MATH SCORE and WRITING SCORE is:	+.43.

This means that the best predictor of FROSH GPA is the MATH SCORE with a correlation of +.77. Adding the other two predictor variables, READING SCORE and WRITING SCORE, improved the prediction by only 0.03 to +.80 and was, therefore, only slightly better in prediction. MATH SCORES are an excellent predictor of FROSH GPA all by themselves.

If you want to learn more about the correlation matrix, see Levine et al. (2011).

7.5   End-of-Chapter Practice Problems

1. The Graduate Record Examinations (GRE) are a standardized test that is an admissions requirement for many US graduate schools. The test is intended to measure general academic preparedness, regardless of specialization field. The general GRE test produces three subtest scores: (1) GRE verbal reasoning (scale 200–800), (2) GRE quantitative reasoning (scale 200–800), and (3) analytical writing (scale 0–6).

    Suppose that you have been asked by the Chair of the Psychology Department at a selective graduate school to see how well the GRE predicts GPA at the end of the first year of graduate study in Psychology. This Chair has asked you to use the three subtest scores of the GRE as predictors and, in addition, to use the GRE Psychology Subject Test score (range 200–800) as an additional predictor of this GPA. The Chair would like your recommendation as to whether or not the Psychology Test should become an admissions requirement in addition to the GRE for admission to the graduate program in psychology. About 7,000 prospective students take the GRE Psychology Test each year.

    You have decided to use a multiple correlation and multiple regression analysis and to test your Excel skills, you have collected the data of a random sample of 12 Psychology students who have just finished their first year of graduate study at this university. These hypothetical data appear in Fig. 7.10:

    (a) Create an Excel spreadsheet using FIRST-YEAR GPA as the criterion (Y), and GRE VERBAL (X_1), GRE QUANTITATIVE (X_2), GRE WRITING (X_3), and GRE PSYCHOLOGY(X_4) as the predictors.
    (b) Use Excel's *multiple regression* function to find the relationship between these five variables and place it below the table.
    (c) Use number format (two decimal places) for the multiple correlation on the SUMMARY OUTPUT and use four decimal places for the coefficients in the SUMMARY OUTPUT.
    (d) Print the table and regression results below the table so that they fit onto one page.
    (e) Save this file as: GRE6.

    Answer the following questions using your Excel printout:

    1. What is the multiple correlation R_{xy}?
    2. What is the y-intercept a?
    3. What is the coefficient for GRE VERBAL b_1?
    4. What is the coefficient for GRE QUANTITATIVE b_2?
    5. What is the coefficient for GRE WRITING b_3?
    6. What is the coefficient for GRE PSYCHOLOGY b_4?
    7. What is the multiple regression equation?
    8. Predict the FIRST-YEAR GPA you would expect for a GRE VERBAL score of 610, a GRE QUANTITATIVE score of 550, a GRE WRITING score of 3, and a GRE PSYCHOLOGY score of 610.

GRADUATE RECORD EXAMINATIONS

How well does the GRE and the GRE subject area test in Psychology predict
GPA at the end of the first year of a Masters' program in Psychology?

FIRST-YEAR GPA	GRE VERBAL	GRE QUANTITATIVE	GRE WRITING	GRE PSYCHOLOGY
3.25	600	620	5	650
3.42	520	550	4	600
2.85	510	540	2	500
2.65	480	460	1	510
3.65	720	710	6	630
3.16	570	610	3	550
3.56	710	650	4	610
2.35	500	480	2	430
2.86	450	470	3	450
2.95	560	530	4	550
3.15	550	580	4	580
3.45	610	620	5	620

Fig. 7.10 Worksheet data for Chap. 7: practice problem #1

(f) Now, go back to your Excel file and create a *correlation matrix* for these five variables, and place it underneath the SUMMARY OUTPUT.

(g) Save this file as: GRE7.

(h) Now, print out *just this correlation matrix* on a separate sheet of paper.

Answer the following questions using your Excel printout. Be sure to include the plus or minus sign for each correlation:

9. What is the correlation GRE VERBAL and FIRST-YEAR GPA?
10. What is the correlation between GRE QUANTITATIVE and FIRST-YEAR GPA?
11. What is the correlation between GRE WRITING and FIRST-YEAR GPA?
12. What is the correlation between GRE PSYCHOLOGY and FIRST-YEAR GPA?
13. What is the correlation between GRE WRITING and GRE VERBAL?
14. What is the correlation between GRE VERBAL and GRE PSYCHOLOGY?
15. Discuss which of the four predictors is the best predictor of FIRST-YEAR GPA.
16. Explain in words how much better the four predictor variables together predict FIRST-YEAR GPA than the best single predictor by itself.

2. The National Football League (2009) and ESPN (2009a, b) record a large number of statistics about players, teams, and leagues on their Web sites. Suppose that you wanted to record the data for 2009 and to create a multiple regression equation for predicting the number of wins during the regular season based on four predictors: (1) yards gained on offense, (2) points scored on offense,

NATIONAL FOOTBALL LEAGUE (NFL)   2009 Regular Season

Team	Games Won	Offense		Defense	
		Yards Gained	Points Scored	Yards allowed	Points allowed
Arizona	10	5510	375	5543	325
Atlanta	9	5447	363	5582	325
Baltimore	9	5619	391	4808	261
Buffalo	6	4382	258	5449	326
Carolina	8	5297	315	5053	308
Chicago	7	4965	327	5404	375
Cincinnati	10	4946	305	4822	291
Cleveland	5	4163	245	6229	375
Dallas	11	6390	361	5054	250
Denver	8	5463	326	5040	324
Detroit	2	4784	262	6274	494
Green Bay	11	6065	461	4551	297
Indianapolis	14	5809	416	5427	307
Jacksonville	7	5385	290	5637	380
Kansas City	4	4851	294	6211	424
Miami	7	5401	360	5589	390
Minnesota	12	6074	470	4888	312
New England	10	6357	427	5123	285
New Orleans	13	6461	510	5724	341
NY Giants	8	5856	402	5198	427
NY Jets	9	5136	348	4037	236
Oakland	5	4258	197	5791	379
Philadelphia	11	5726	429	5137	337
Pittsburgh	9	5941	368	4885	324
San Diego	13	5761	454	5230	320
San Francisco	8	4652	330	5222	281
Seattle	5	5069	280	5703	390
St. Louis	1	4470	175	5965	436
Tempa Bay	3	4600	244	5849	400
Tennesee	8	5623	354	5850	402
Washington	4	4998	266	5115	336
Houston	9	6129	388	5198	333

Fig. 7.11 Worksheet data for Chap. 7: practice problem #2

(3) yards allowed on defense, and (4) points allowed on defense. These data are recorded in the table in Fig. 7.11:

(a) Create an Excel spreadsheet using games won as the criterion (Y), and the other variables as the four predictors of this criterion.

(b) Use Excel's *multiple regression* function to find the relationship between these variables and place it below the table.

(c) Use number format (two decimal places) for the multiple correlation on the SUMMARY OUTPUT, and use number format (three decimal places) for the coefficients in the SUMMARY OUTPUT.

(d) Print the table and regression results below the table so that they fit onto one page.

(e) By hand on this printout, *circle and label.*

(1a) Multiple correlation R_{xy}.

(2b) Coefficients for the y-intercept, yards gained, points scored, yards allowed, and points allowed.

(f) Save this file as: NFL2009B.

(g) Now, go back to your Excel file and create a correlation matrix for these five variables, and place it underneath the Summary Table. *Change each correlation to just two decimals.* Save this file as NFL2009C.

(h) Now, print out *just this correlation matrix in portrait mode* on a separate sheet of paper.

Answer the following questions using your Excel printout:

1. What is the multiple correlation R_{xy}?
2. What is the y-intercept a?
3. What is the coefficient for yards gained b_1?
4. What is the coefficient for points scored b_2?
5. What is the coefficient for yards allowed b_3?
6. What is the coefficient for points allowed b_4?
7. What is the multiple regression equation?
8. Underneath this regression equation by hand, predict the number of wins you would expect for 5,100 yards gained, 360 points scored, 5,400 yards allowed, and 330 points allowed.

Answer the following questions using your Excel printout. Be sure to include the plus or minus sign for each correlation:

9. What is the correlation between yards gained and games won?
10. What is the correlation between points scored and games won?
11. What is the correlation between yards allowed and games won?
12. What is the correlation between points allowed and games won?
13. What is the correlation between points scored and yards gained?
14. What is the correlation between points allowed and points scored?
15. Discuss which of the four predictors is the best predictor of games won.
16. Explain in words how much better the four predictor variables combined predict games won than the best single predictor by itself.

3. Suppose that you are the marketing manager for 7-Eleven stores in Missouri and that you want to see if a proposed store location would generate sufficient yearly sales volume to support the idea of building a new store at that location. You have checked the data available at your company to generate the following table for a random sample of 20 7-Eleven stores in Missouri based on last year's data to create the hypothetical data given in Fig. 7.12:

(a) Create an Excel spreadsheet using the annual sales figures as the criterion and the average daily traffic, population, and income figures as the predictors.

7Eleven Stores in Missouri

Store ID	Y Annual Sales ($000)	X₁ Average Daily Traffic	X₂ Population (2-mile radius)	X₃ Average Income in Area
1	1,121	61,655	17,880	$28,991
2	766	35,236	13,742	$14,731
3	595	35,403	19,741	$8,114
4	899	52,832	23,246	$15,324
5	915	40,809	24,485	$11,438
6	782	40,820	20,410	$11,730
7	833	49,147	28,997	$10,589
8	571	24,953	9,981	$10,706
9	692	40,828	8,982	$23,591
10	1,005	39,195	18,814	$15,703
11	589	34,574	16,941	$9,015
12	671	26,639	13,319	$10,065
13	903	55,083	21,482	$17,365
14	703	37,892	26,524	$7,532
15	556	24,019	14,412	$6,950
16	657	27,791	13,896	$9,855
17	1,209	53,438	22,444	$21,589
18	997	54,835	18,096	$22,659
19	844	32,919	16,458	$12,660
20	883	29,139	16,609	$11,618

Fig. 7.12 Worksheet data for Chap. 7: practice problem #3

(b) Use Excel's *multiple regression* function to find the relationship between these four variables and place the SUMMARY OUTPUT below the table.

(c) Use number format (two decimal places) for the multiple correlation on the SUMMARY OUTPUT, and use this same number format for the coefficients in the SUMMARY OUTPUT.

(d) Save the file as: multiple2.

(e) Print the table and regression results below the table so that they fit onto one page.

Answer the following questions using your Excel printout:

1. What is multiple correlation R_{xy}?
2. What is the y-intercept a?
3. What is the coefficient for average daily traffic b_1?
4. What is the coefficient for population b_2?
5. What is the coefficient for average income b_3?
6. What is the multiple regression equation?
7. Predict the annual sales you would expect for average daily traffic of 42,000, a population of 23,000, and income of $22,000.

(f) Now, go back to your Excel file and create a correlation matrix for these four variables and place it underneath the SUMMARY OUTPUT on your spreadsheet.

(g) Save this file as: multiple3.

(h) Now, print out *just this correlation matrix* on a separate sheet of paper.

Answer the following questions using your Excel printout. Be sure to include the plus or minus sign for each correlation:

8. What is the correlation between traffic and sales?
9. What is the correlation between population and sales?
10. What is the correlation between income and sales?
11. What is the correlation between traffic and population?
12. What is the correlation between population and income?
13. Discuss which of the three predictors is the best predictor of annual sales.
14. Explain in words how much better the three predictor variables combined predict annual sales than the best single predictor by itself.

References

ESPN. NFL Team Total Offense Statistics – 2009a. Retrieved December 9, 2010, from http://espn.go.com/nfl/statistics/team/_/stat/total/year/2009

ESPN. NFL Team Total Defense Statistics – 2009b. Retrieved December 9, 2010, from http://espn.go.com/nfl/statistics/team/_/stat/total/position/defense/year/2009

Keller, G. Statistics for Management and Economics (8th ed.). Mason, OH: South-Western Cengage Learning, 2009.

Levine, D.M., Stephan, D.F., Krehbiel, T.C., and Berenson, M.L. Statistics for Managers using Microsoft Excel (6th ed.). Boston, MA: Prentice Hall/Pearson, 2011.

National Football League. Standings [2009 Regular Season by league]. Retrieved December 9, 2010 http://www.nfl.com/standings?category=league&season=2009-REG&split=Overall

Chapter 8
One-Way Analysis of Variance (ANOVA)

So far in this 2010 Excel Guide, you have learned how to use a one-group t-test to compare the sample mean to the population mean and a two-group t-test to test for the difference between two sample means. *But what should you do when you have more than two groups and you want to determine if there is a significant difference between the means of these groups?*

The answer to this question is: *Analysis of Variance (ANOVA).*

The ANOVA test allows you to test for the difference between the means when you have *three or more groups* in your research study.

Important note: *In order to do one-way analysis of variance, you need to have installed the "Data Analysis Toolpak" that was described in Chap. 6 (see Sect. 6.5.1). If you did not install this, you need to do that now.*

Let us suppose that an undergraduate introductory statistics course in the Education Department at a State University was taught for the spring semester in three different formats: (1) computer-assisted instruction (CAI), (2) in-class lectures (LECTURES), and (3) independent study in an online version using a standard textbook (INDEPENDENT). Suppose, further, that all students took the same comprehensive final examination (100 points possible).

You have been asked to analyze the data from the final examinations to determine if there was any significant difference in final exam scores between the three types of teaching methods. To test your Excel skills, you have selected a random sample of students from each of these methods (see Fig. 8.1). Note that each group of students can be of a different number of students in order for ANOVA to be used on the data. Statisticians delight in this fact by referring to this characteristic by stating that: "ANOVA is a very robust test." (Statisticians love that term!)

T. Quirk, *Excel 2010 for Educational and Psychological Statistics:*
A Guide to Solving Practical Problems, DOI 10.1007/978-1-4614-2071-2_8,
© Springer Science+Business Media, LLC 2012

	A	B	C	D	
1					
2	STATE UNIVERSITY				
3					
4	UNDERGRADUATE BASIC STATISTICS COURSE: FINAL EXAM				
5					
6	CAI	LECTURES	INDEPENDENT		
7	90	85	76		
8	85	89	80		
9	74	83	90		
10	89	79	84		
11	84	74	78		
12	95	75	65		
13	92	86	42		
14	65	87	58		
15	75	86	63		
16	73	88	75		
17	54		66		
18	71				
19					
20					

Fig. 8.1 Worksheet data for statistics teaching methods (practical example)

Create an Excel spreadsheet for these data in this way:

A2:   STATE UNIVERSITY
A4:   UNDERGRADUATE BASIC STATISTICS COURSE: FINAL EXAM
A6:   CAI
A7:   90
B6:   LECTURES
C6:   INDEPENDENT
A18:   71

Enter the other information into your spreadsheet table. When you have finished entering these data, the last cell on the left should have 71 in cell A18, and the last cell on the right should have 66 in cell C17. Center the numbers in each of the columns. Use number format (zero decimals) for all numbers.

Important note: *Be sure to double-check all of your figures in the table to make sure that they are exactly correct or you will not be able to obtain the correct answer for this problem!*

Save this file as: STATS3.

8.1   Using Excel to Perform a One-Way Analysis of Variance (ANOVA)

Objective: To use Excel to perform a one-way ANOVA test

You are now ready to perform an ANOVA test on these data using the following steps:

Data (at the top of screen)
Data analysis (far right at the top of screen)
ANOVA: Single Factor (*scroll up to this formula and highlight it*; see Fig. 8.2)

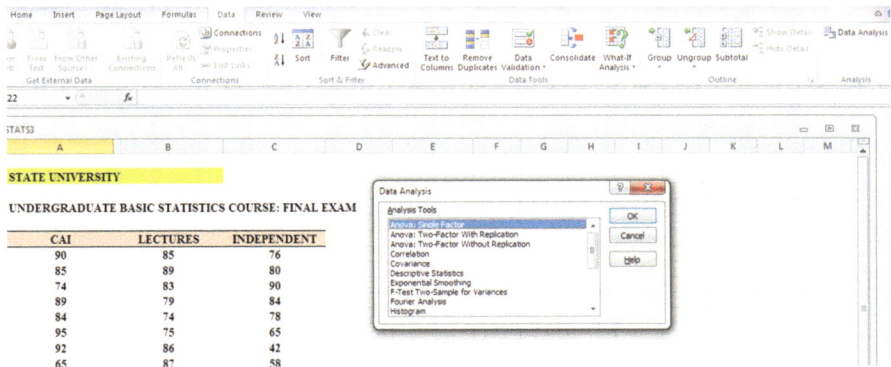

Fig. 8.2 Dialog box for data analysis: ANOVA Single Factor

OK
Input range: A6:C18 (note that you have included in this range the column titles that are in row 6).

Important note: *Whenever the data set has a different sample size in the groups being compared, the INPUT RANGE that you define must start at the column title of the first group on the left and go to the last column on the right to the lowest row that has a figure in it in the entire data matrix so that the INPUT RANGE has the "shape" of a rectangle when you highlight it.*

Grouped by: columns.
Put a check mark in: labels in first row.
Output range (click on the button to its left):   A20 (see Fig. 8.3)

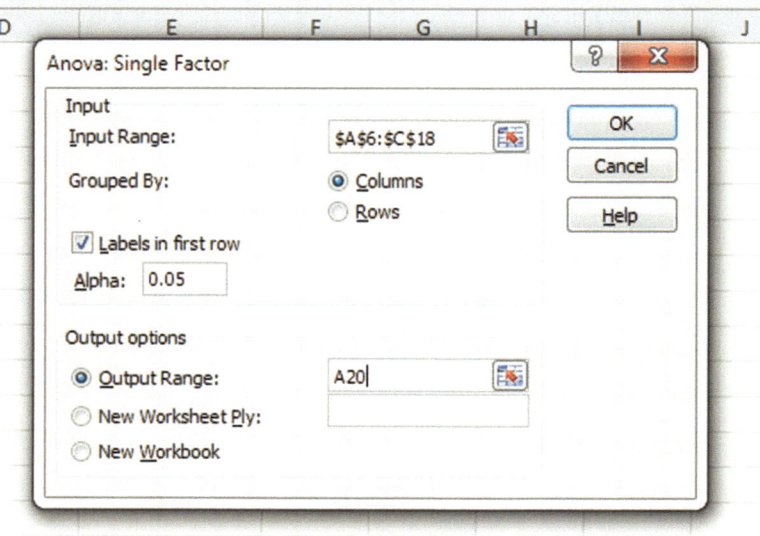

Fig. 8.3 Dialog box for ANOVA: Single Factor Input/Output Range

OK.

Next, format all decimal figures in the SUMMARY to two decimal places as this
will make the printout much easier to read. Center all figures in their cells.

Save this file as: STATS4.

You should have generated the table given in Fig. 8.4.

	A	B	C	D	E	F	G	H
17	54		66					
18	71							
19								
20	**Anova: Single Factor**							
21								
22	**SUMMARY**							
23	*Groups*	*Count*	*Sum*	*Average*	*Variance*			
24	**CAI**	12	947	78.92	151.72			
25	**LECTURES**	10	832	83.20	28.84			
26	**INDEPENDENT**	11	777	70.64	183.45			
27								
28								
29	**ANOVA**							
30	*Source of Variation*	*SS*	*df*	*MS*	*F*	*P-value*	*F crit*	
31	**Between Groups**	867.12	2	433.56	3.46	0.04	3.32	
32	**Within Groups**	3763.06	30	125.44				
33								
34	Total	4630.18	32					
35								
36								

Fig. 8.4 ANOVA results for statistics teaching methods

Print out both the data table and the ANOVA summary table so that all of this information fits onto one page. (Hint: Set the Page Layout/Fit to Scale to *80% size*).

As a check on your analysis, you should have the following in these cells:

A20: ANOVA: Single Factor
D31: 433.56
G31: 3.32
D24: 78.92

Now, let us discuss how you should interpret this table.

8.2 How to Interpret the ANOVA Table Correctly

> Objective: To interpret the ANOVA table correctly

ANOVA allows you to test for the differences between means when you have three or more groups of data. This ANOVA test is called the *F*-test statistic and is typically identified with the letter *F*.

The formula for the *F*-test is this:

F = Mean square between groups (MS_b) divided by mean square within groups (MS_w)

$$F = MS_b/MS_w \qquad (8.1)$$

The derivation and explanation of this formula is beyond the scope of this *Excel Guide*. In this *Excel Guide*, we are attempting to teach you *how to use Excel*, and we are not attempting to teach you the statistical theory that is behind the ANOVA formulas. For a detailed explanation of ANOVA, see Weiers (2011).

Note that cell D31 contains $MS_b = 433.56$, while cell D32 contains $MS_w = 125.44$.

When you divide these two figures using their cell references in Excel, you get the answer for the *F*-test of 3.46 which is in cell E31. Let us discuss now the meaning of the figure: $F = 3.46$.

In order to determine whether this figure for *F* of 3.46 indicates a significant difference between the means of the three groups, the first step is to write the null hypothesis and the research hypothesis for the three teaching formats.

In our statistics teaching methods comparisons, the null hypothesis states that the population means of the three groups are equal, while the research hypothesis states that the population means of the three groups are not equal and that there is, therefore, a significant difference between the population means of the three groups. Which of these two hypotheses should you accept based on the ANOVA results?

8.3   Using the Decision Rule for the ANOVA *F*-Test

To state the hypotheses, let us call CAI as group 1, LECTURES as group 2, and INDEPENDENT as group 3. The hypotheses would then be:

$$H_0 : \mu_1 = \mu_2 = \mu_3$$

$$H_1 : \mu_1 \neq \mu_2 \neq \mu_3$$

The answer to this question is analogous to the decision rule used in this book for both the one-group t-test and the two-group t-test. You will recall that this rule (see Sects. 4.1.6 and 5.1.8) was:

If the absolute value of t is less than the critical t, you accept the null hypothesis.
or
If the absolute value of t is greater than the critical t, you reject the null hypothesis and accept the research hypothesis.

Now, here is the decision rule for ANOVA:

Objective: To learn the decision rule for the ANOVA *F*-test

The decision rule for the ANOVA *F*-test is the following:

If the value for F is less than the critical F-value, accept the null hypothesis.
or
If the value of F is greater than the critical F-value, reject the null hypothesis and accept the research hypothesis.

Note that Excel tells you the critical *F*-value in cell G31: 3.32
Therefore, our decision rule for the teaching methods AVOVA test is this:

Since the value of F of 3.46 is greater than the critical F-value of 3.32, we reject the null hypothesis and accept the research hypothesis.

Therefore, our conclusion, in plain English, is:

There is a significant difference between the final exam scores in the three types of teaching methods.

Note that it is not necessary to take the absolute value of *F* of 3.46. The *F*-value can never be less than one, and so it can never be a negative value which requires us to take its absolute value in order to treat it as a positive value.

It is important to note that ANOVA tells us that there was a significant difference between the population means of the three groups, *but it does not tell us which pairs of groups were significantly different from each other.*

8.4   Testing the Difference Between Two Groups Using the ANOVA *t*-Test

To answer that question, we need to do a different test called the ANOVA *t*-test.

> Objective: To test the difference between the means of two groups using an ANOVA *t*-test when the ANOVA results indicate a significant difference between the population means.

Since we have three groups of data (one group for each of the three teaching methods), we would have to perform three separate ANOVA *t*-tests to determine which pairs of groups were significantly different. This means that we would have to perform a separate ANOVA *t*-test for the following pairs of groups:

1. CAI vs. LECTURES
2. CAI vs. INDEPENDENT
3. LECTURES vs. INDEPENDENT

We will do just one of these pairs of tests, LECTURES vs. INDEPENDENT, to illustrate the way to perform an ANOVA *t*-test comparing these two teaching methods. The ANOVA *t*-test for the other two pairs of groups would be done in the same way.

8.4.1   Comparing LECTURES vs. INDEPENDENT in Their Exam Scores Using the ANOVA t-Test

> Objective: To compare LECTURES vs. INDEPENDENT in their final exam scores using the ANOVA *t*-test.

The first step is to write the null hypothesis and the research hypothesis for these two teaching methods.

For the ANOVA *t*-test, the null hypothesis is that the population means of the two groups are equal, while the research hypothesis is that the population means of the two groups are not equal (i.e., there is a significant difference between these two means). Since we are comparing LECTURES (group 2) vs. INDEPENDENT (group 3), these hypotheses would be:

$$H_0 : \mu_2 = \mu_3$$

$$H_1 : \mu_2 \neq \mu_3$$

For group 2 vs. group 3, the formula for the ANOVA t-test is:

$$\text{ANOVA } t = \frac{\bar{X}_1 - \bar{X}_2}{\text{s.e.}_{\text{ANOVA}}} \tag{8.2}$$

Where

$$\text{s.e.}_{\text{ANOVA}} = \sqrt{\text{MS}_w \left(\frac{1}{n_1} + \frac{1}{n_2} \right)} \tag{8.3}$$

The steps involved in computing this ANOVA t-test are:

1. Find the difference of the sample means for the two groups $(83.20 - 70.64 = 12.56)$.
2. Find $1/n_2 + 1/n_3$ (since both groups have a different number of students in them, this becomes $1/10 + 1/11 = 0.10 + 0.09 = 0.19$).
3. Multiply MS_w times the answer for step 2 ($125.44 \times 0.19 = 23.83$).
4. Take the square root of step 3 (SQRT $(23.83) = 4.88$).
5. Divide step 1 by step 4 to find ANOVA t ($12.56/4.88 = 2.57$).

Note: Since Excel computes all calculations to 16 decimal places, when you use Excel for the above computations, your answer will be 2.57 in two decimal places, but Excel's answer will be much more accurate because it is always in 16 decimal places.

Now, what do we do with this ANOVA t-test result of 2.57? In order to interpret this value of 2.57 correctly, we need to determine the critical value of t for the ANOVA t-test. To do that, we need to find the degrees of freedom for the ANOVA t-test as follows:

8.4.1.1 Finding the Degrees of Freedom for the ANOVA t-Test

Objective: To find the degrees of freedom for the ANOVA t-test

The degrees of freedom (df) for the ANOVA t-test is found as follows:

df = take the total sample size of all of the groups and subtract the number of groups in your study ($n_{\text{TOTAL}} - k$ where k = the number of groups)

In our example, the total sample size of the three groups is 33 since there are 12 students in group1, 10 students in group 2, and 11 students in group 3, and since there are three groups, $33 - 3$ gives a degrees of freedom for the ANOVA t-test of 30.

If you look up df = 30 in the t-table in Appendix E in the degrees of freedom column (df), which is the *second column on the left of this table*, you will find that the critical t-value is 2.042.

Important note: *Be sure to use the degrees of freedom column (df) in Appendix E for the ANOVA t-test critical t value*

8.4.1.2 Stating the Decision Rule for the ANOVA *t*-test

Objective: To learn the decision rule for the ANOVA *t*-test

Interpreting the result of the ANOVA *t*-test follows the same decision rule that we used for both the one-group *t*-test (see Sect. 4.1.6) and the two-group *t*-test (see Sect. 5.1.8):

If the absolute value of t is less than the critical value of t, we accept the null hypothesis.

or

If the absolute value of t is greater than the critical value of t, we reject the null hypothesis and accept the research hypothesis.

Since we are using a type of *t*-test, we need to take the absolute value of *t*. Since the absolute value of 2.57 is greater than the critical *t*-value of 2.042, we reject the null hypothesis (that the population means of the two groups are equal) and accept the research hypothesis (that the population means of the two groups are significantly different from one another).

This means that our conclusion, in plain English, is as follows:

The average final exam scores for the LECTURES group were significantly higher than the average scores for the INDEPENDENT group (83.20 vs. 70.64).

Note that this difference in average exam scores of almost 13 points might not seem like much, but in practical terms, this means that the average scores in the LECTURES group are 18% higher than the average scores in the INDEPENDENT group. This, clearly, is an important difference in final exam scores based on our hypothetical data.

8.4.1.3 Performing an ANOVA *t*-Test Using Excel Commands

Now, let us do these calculations for the ANOVA *t*-test using Excel with the file you created earlier in this chapter: STATS4

A37:   LECTURES vs. INDEPENDENT STUDY
A39: $1/n$ LECTURES + $1/n$ INDEPENDENT
A41:   s.e. of LECTURES vs. INDEPENDENT
A43:   ANOVA *t*-test
C39:   =(1/10 + 1/11)
C41:   =SQRT(D32*C39)
C43:   =(D25 − D26)/C41

You should now have the following results in these cells when you round off all these figures in the ANOVA *t*-test to two decimal points:

C39:   0.19
C41:   4.89
C43:   2.57

Save this final result under the file name STATS5.

Print out the resulting spreadsheet so that it fits onto one page like Fig. 8.5 (Hint: Reduce the Page Layout/Scale to Fit to *80%*).

For a more detailed explanation of the ANOVA *t*-test, see Black (2010).

STATE UNIVERSITY

UNDERGRADUATE BASIC STATISTICS COURSE: FINAL EXAM

CAI	LECTURES	INDEPENDENT
90	85	76
85	89	80
74	83	90
89	79	84
84	74	78
95	75	65
92	86	42
65	87	58
75	86	63
73	88	75
54		66
71		

Anova: Single Factor

SUMMARY

Groups	Count	Sum	Average	Variance
CAI	12	947	78.92	151.72
LECTURES	10	832	83.20	28.84
INDEPENDENT	11	777	70.64	183.45

ANOVA

Source of Variation	SS	df	MS	F	P-value	F crit
Between Groups	867.12	2	433.56	3.46	0.04	3.32
Within Groups	3763.06	30	125.44			
Total	4630.18	32				

LECTURES vs. INDEPENDENT STUDY

1/n LECTURES + 1/n INDEPENDENT	0.19
s.e. of LECTURES vs. INDEPENDENT	4.89
ANOVA t-test	2.57

Fig. 8.5 Final spreadsheet of final exam scores for LECTURES vs. INDEPENDENT groups

Important note: *You are only allowed to perform an ANOVA t-test comparing the population means of two groups when the F-test produces a significant difference between the population means of all of the groups in your study.*

   It is improper to do any ANOVA t-test when the value of F is less than the critical value of F. Whenever F is less than the critical F, this means that there was no difference between the population means of the groups and, therefore, that you cannot test to see if there is a difference between the means of any two groups since this would capitalize on chance differences between these two groups.

8.5   End-of-Chapter Practice Problems

1. The Iowa Tests of Basic Skills (ITBS) are a standardized achievement test battery that is used in grades K-8 to assess the readiness of pupils in terms of their reading comprehension ability, as well as other areas of achievement. One type of score generated from these tests is the grade equivalent score (GE) which attempts to describe the reading comprehension ability of the student in terms of grade level. A GE score of 6.4, for a fifth grader, for example, means that this pupil is reading at the same level as the typical sixth grader finishing the fourth month of the school year in sixth grade. GE scores attempt to measure a pupil's developmental level only and are not intended to suggest the grade level to which the pupil should be placed.

   Suppose that you have been asked to analyze the GE scores of fifth graders in three different elementary schools in the same school district. These schools have historically been very similar in the parents' educational level, household income, and other important demographic factors. Suppose that each school has been using a different method of teaching reading during the fifth grade and that all fifth grade students in these schools took the ITBS in May of their fifth grade.

   To test your Excel skills, you have randomly selected students from the fifth grade in these three schools and recorded their GE scores given in the hypothetical data of Fig. 8.6.

IOWA TESTS OF BASIC SKILLS

GRADE 5 READING COMPREHENSION SCORES

GRADE EQUIVALENT SCORES (GE)

SCHOOL A	SCHOOL B	SCHOOL C
5.2	5.1	5.8
5.5	5.6	5.6
5.9	5.8	6.3
4.6	6.4	5.9
6.2	6.5	6.4
6.4	6.7	6.1
5.8	4.8	6.4
5.9	4.9	6.5
5.4	4.3	6.8
6.3	6.5	7.2
6.4	6.2	6.9
7.3		7.3
		7.4

Fig. 8.6 Worksheet data for Chap. 8: practice problem #1

(a) Enter these data on an Excel spreadsheet.
(b) Perform a *one-way ANOVA test* on these data and show the resulting ANOVA table *underneath* the input data for the three schools.
(c) If the F-value in the ANOVA table is significant, create an Excel formula to compute the ANOVA t-test comparing the average for SCHOOL A against SCHOOL C and show the results below the ANOVA table on the spreadsheet (put the standard error and the ANOVA t-test value on separate lines of your spreadsheet and use two decimal places for each value).
(d) Print out the resulting spreadsheet so that all of the information fits onto one page.
(e) Save the spreadsheet as READING5.

Now, write the answers to the following questions using your Excel printout:

1. What are the null hypothesis and the research hypothesis for the ANOVA F-test?
2. What is MS_b on your Excel printout?
3. What is MS_w on your Excel printout?
4. Compute $F = MS_b/MS_w$ using your calculator.
5. What is the critical value of F on your Excel printout?
6. What is the result of the ANOVA F-test?
7. What is the conclusion of the ANOVA F-test in plain English?
8. If the ANOVA F-test produced a significant difference between the three schools in reading comprehension GE scores, what is the null hypothesis and the research hypothesis for the ANOVA t-test comparing SCHOOL A vs. SCHOOL C?

9. What is the GE mean (average) for SCHOOL A on your Excel printout?
10. What is the GE mean (average) for SCHOOL C on your Excel printout?
11. What are the degrees of freedom (df) for the ANOVA t-test comparing SCHOOL A vs. SCHOOL C?
12. What is the critical t value for this ANOVA t-test in Appendix E for these degrees of freedom?
13. Compute the s.e.$_{ANOVA}$ using your calculator.
14. Compute the ANOVA t-test value comparing SCHOOL A vs. SCHOOL C using your calculator.
15. What is the result of the ANOVA t-test comparing SCHOOL A vs. SCHOOL C?
16. What is the conclusion of the ANOVA t-test comparing SCHOOL A vs. SCHOOL C in plain English?
    Note that since there are three schools, you need to do three ANOVA t-tests to determine what the significant differences are between the schools. *Since you have just completed the ANOVA t-test comparing SCHOOL A vs. SCHOOL C, let us do the ANOVA t-test next comparing SCHOOL A vs. SCHOOL B.*
17. State the null hypothesis and the research hypothesis comparing SCHOOL A vs. SCHOOL B.
18. What is the GE mean (average) for SCHOOL A on your Excel printout?
19. What is the GE mean (average) for SCHOOL B on your Excel printout?
20. What are the degrees of freedom (df) for the ANOVA t-test comparing SCHOOL A vs. SCHOOL B?
21. What is the critical t value for this ANOVA t-test in Appendix E for these degrees of freedom?
22. Compute the s.e.$_{ANOVA}$ for SCHOOL A vs. SCHOOL B using your calculator.
23. Compute the ANOVA t-test value comparing SCHOOL A vs. SCHOOL B.
24. What is the result of the ANOVA t-test comparing SCHOOL A vs. SCHOOL B?
25. What is the conclusion of the ANOVA t-test comparing SCHOOL A vs. SCHOOL B in plain English?
    The last ANOVA t-test compares SCHOOL B vs. SCHOOL C. Let us do that test below:
26. State the null hypothesis and the research hypothesis comparing SCHOOL B vs. SCHOOL C.
27. What is the GE mean (average) for SCHOOL B on your Excel printout?
28. What is the GE mean (average) for SCHOOL C on your Excel printout?
29. What are the degrees of freedom (df) for the ANOVA t-test comparing SCHOOL B vs. SCHOOL C?
30. What is the critical t value for this ANOVA t-test in Appendix E for these degrees of freedom?
31. Compute the s.e.$_{ANOVA}$ comparing SCHOOL B vs. SCHOOL C using your calculator.

32. Compute the ANOVA t-test value comparing SCHOOL B vs. SCHOOL C with your calculator.
33. What is the result of the ANOVA t-test comparing SCHOOL B vs. SCHOOL C?
34. What is the conclusion of the ANOVA t-test comparing SCHOOL B vs. SCHOOL C in plain English?
35. What is the summary of the three ANOVA t-tests in plain English?
36. What recommendation would you make to school superintendent about these three types of reading instruction based on the results of your analysis? Why would you make that recommendation?

2. McDonald's rolled out the "100% Angus Beef Third Pounders Burgers" in July 2009 to compete with the supersize hamburgers sold by Hardee's. Suppose that you had been hired as a consultant by McDonald's to analyze the data from a test market study involving four test market cities matched for population size, average household income, average family size, and number of McDonald's restaurants in each city. Suppose, further, that the test market ran for 12 weeks, and that each city used only one type of advertisement for these burgers: (1) Radio, (2) Local TV, (3) Billboards, and (4) Local newspaper. The cities were randomly assigned to one type of ad, and each city spent the same advertising dollars each week on their one type of ad. The hypothetical data for the number of units sold each week of the Angus Burger are given in Fig. 8.7.

(a) Enter these data on an Excel spreadsheet.

ANGUS BURGER TEST MARKET STUDY

1	2	3	4
Radio	Local TV	Billboards	Local newspaper
300	310	340	280
320	315	330	285
310	320	345	290
290	326	342	275
280	324	341	282
315	318	351	284
326	330	339	291
295	327	337	284
278	328	329	279
289	319	328	274
287	326	332	283
305	328	335	285

Fig. 8.7 Worksheet data for Chap. 8: practice problem #2

(b) Perform a *one-way ANOVA test* on these data and show the resulting ANOVA table *underneath* the input data for the four types of ads.

(c) If the *F*-value in the ANOVA table is significant, create an Excel formula to compute the ANOVA *t*-test comparing the average number of units sold for Billboard ads against the average for Radio ads and show the results below the ANOVA table on the spreadsheet (put the standard error and the ANOVA *t*-test value on separate lines of your spreadsheet and use two decimal places for each value)

(d) Print out the resulting spreadsheet so that all of the information fits onto one page

(e) Save the spreadsheet as McD4.

Let us call the Radio ads group 1, the Local TV ads group 2, the Billboards ads group 3, and the Local Newspaper ads group 4.

Now, write the answers to the following questions using your Excel printout:

1. What are the null hypothesis and the research hypothesis for the ANOVA *F*-test?
2. What is MS_b on your Excel printout?
3. What is MS_w on your Excel printout?
4. Compute $F = MS_b/MS_w$ using your calculator.
5. What is the critical value of *F* on your Excel printout?
6. What is the result of the ANOVA *F*-test?
7. What is the conclusion of the ANOVA *F*-test in plain English?
8. If the ANOVA *F*-test produced a significant difference between the four types of ads in the number of Angus Burgers sold per week, what is the null hypothesis and the research hypothesis for the ANOVA *t*-test comparing Billboards ads (group 3) vs. Radio ads (group 1)?
9. What is the mean (average) for Billboards ads on your Excel printout?
10. What is the mean (average) for Radio ads on your Excel printout?
11. What are the degrees of freedom (df) for the ANOVA *t*-test comparing Billboards ads vs. Radio ads?
12. What is the critical *t* value for this ANOVA *t*-test in Appendix E for these degrees of freedom?
13. Compute the s.e.$_{ANOVA}$ using your calculator for Billboards ads vs. Radio ads.
14. Compute the ANOVA *t*-test value comparing Billboard ads vs. Radio ads using your calculator.
15. What is the result of the ANOVA *t*-test comparing Billboards ads vs. Radio ads?
16. What is the conclusion of the ANOVA *t*-test comparing Billboards ads vs. Radio ads in plain English?

3. Suppose that you have been hired as a consultant by Procter & Gamble to analyze the data from a pilot study involving three recent focus groups who were shown four different television commercials for a new type of Crest toothpaste that have not yet been shown on television. The participants were given a

10-item survey to complete after seeing the commercials, and the hypothetical data from question #8 is given in Fig. 8.8 for the four TV commercials.

ITEM #8:	"How believable is this commercial to you?"

1	2	3	4	5	6	7	8	9
not very believable								very believable

Rating for Focus Groups 1, 2, 3 combined

	Television commercial		
A	B	C	D
2	3	5	6
3	4	6	7
5	5	7	4
4	2	5	5
5	6	8	3
3	1	6	8
6	4	7	2
4	3	5	6
3	7	4	7
7	6	6	5
2	5	3	8
1	3	6	9
3	4	8	5
5	2	9	6
6	3	5	7

Fig. 8.8 Worksheet data for Chap. 8: practice problem #3

(a) Enter these data on an Excel spreadsheet.
(b) Perform a *one-way ANOVA test* on these data and show the resulting ANOVA table *underneath* the input data for the four types of commercials.

(c) If the F-value in the ANOVA table is significant, create an Excel formula to compute the ANOVA t-test comparing the average for Commercial B against the average for Commercial D and show the results below the ANOVA table on the spreadsheet (put the standard error and the ANOVA t-test value on separate lines of your spreadsheet and use two decimal places for each value)

(d) Print out the resulting spreadsheet so that all of the information fits onto one page

(e) Save the spreadsheet as TV6.

Now, write the answers to the following questions using your Excel printout:

1. What are the null hypothesis and the research hypothesis for the ANOVA F-test?
2. What is MS_b on your Excel printout?
3. What is MS_w on your Excel printout?
4. Compute $F = MS_b/MS_w$ using your calculator.
5. What is the critical value of F on your Excel printout?
6. What is the result of the ANOVA F-test?
7. What is the conclusion of the ANOVA F-test in plain English?
8. If the ANOVA F-test produced a significant difference between the four types of TV commercials in their believability, what is the null hypothesis and the research hypothesis for the ANOVA t-test comparing Commercial B vs. Commercial D?
9. What is the mean (average) for Commercial B on your Excel printout?
10. What is the mean (average) for Commercial D on your Excel printout?
11. What are the degrees of freedom (df) for the ANOVA t-test comparing Commercial B vs. Commercial D?
12. What is the critical t value for this ANOVA t-test in Appendix E for these degrees of freedom?
13. Compute the s.e.$_{ANOVA}$ using your calculator for Commercial B vs. Commercial D.
14. Compute the ANOVA t-test value comparing Commercial B vs. Commercial D using your calculator.
15. What is the result of the ANOVA t-test comparing Commercial B vs. Commercial D?
16. What is the conclusion of the ANOVA t-test comparing Commercial B vs. Commercial D in plain English?

References

Black, K. Business Statistics: For Contemporary Decision Making (6th ed.). Hoboken, NJ: John Wiley & Sons, Inc., 2010.
Weiers, R.M. Introduction to Business Statistics (7th ed.). Mason, OH: South-Western Cengage Learning, 2011.

Appendices

Appendix A: Answers to End-of-Chapter Practice Problems

T. Quirk, *Excel 2010 for Educational and Psychological Statistics:*
A Guide to Solving Practical Problems, DOI 10.1007/978-1-4614-2071-2,
© Springer Science+Business Media, LLC 2012

Chapter 1: Practice problem #1 answer (see Fig. A.1)

Zach White School
4th grade unit on *Charlotte's Web*

Chapters 1-5 test (25 items)		
20		
18	n	13
19		
16		
23	MEAN	15.85
24		
10		
12	STDEV	4.90
11		
8		
16	s.e.	1.36
14		
15		

Fig. A.1 Answer to Chap. 1: practice problem #1

Chapter 1: Practice problem #2 answer (see Fig. A.2)

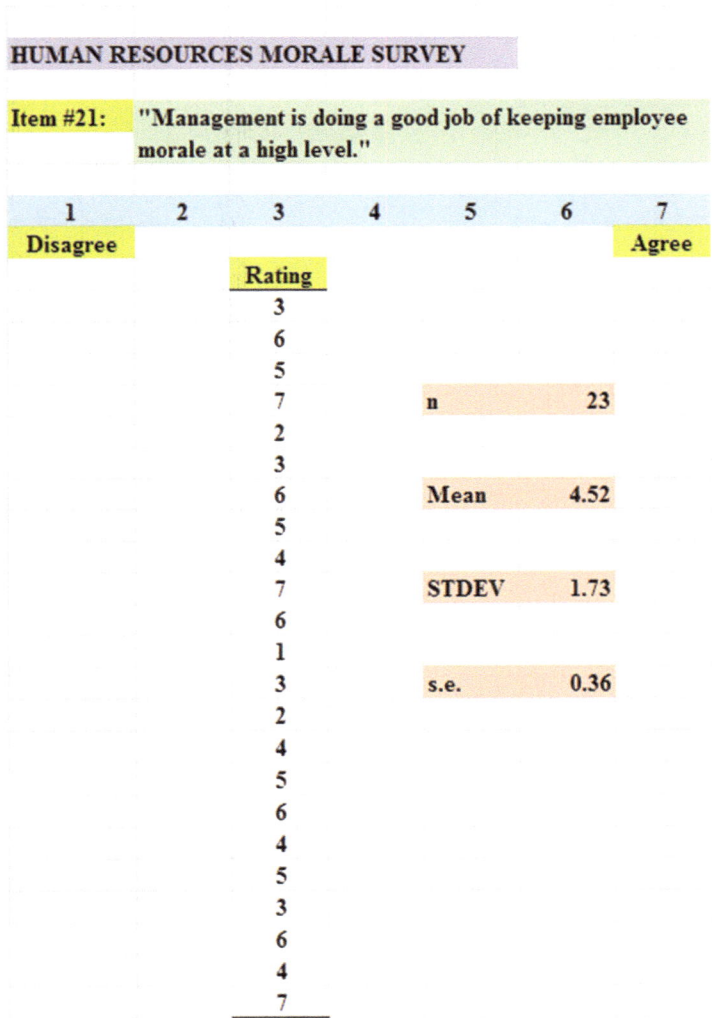

HUMAN RESOURCES MORALE SURVEY

Item #21: "Management is doing a good job of keeping employee morale at a high level."

1	2	3	4	5	6	7
Disagree						Agree

Rating		
3		
6		
5		
7	n	23
2		
3		
6	Mean	4.52
5		
4		
7	STDEV	1.73
6		
1		
3	s.e.	0.36
2		
4		
5		
6		
4		
5		
3		
6		
4		
7		

Fig. A.2 Answer to Chap. 1: practice problem #2

Chapter 1: Practice problem #3 answer (see Fig. A.3)

Deer Creek Elementary School		
5th grade science test		

Chapter 8 (15 items)		
12		
15	n	16
13		
8		
10	MEAN	11.750
12		
13		
12	STDEV	2.910
9		
4		
11	s.e.	0.727
15		
13		
15		
12		
14		

Fig. A.3 Answer to Chap. 1: practice problem #3

Chapter 2: Practice problem #1 answer (see Fig. A.4)

FRAME NUMBERS	Duplicate frame numbers	RANDOM NO.
1	44	0.029
2	33	0.734
3	38	0.885
4	43	0.052
5	13	0.739
6	10	0.574
7	50	0.195
8	1	0.388
9	48	0.584
10	61	0.260
11	4	0.796
12	22	0.234
13	40	0.293
14	37	0.185
15	35	0.484
16	60	0.856
17	59	0.738
18	7	0.394
19	17	1.000
20	39	0.720
21		
	11	0.c
56	56	0.434
57	57	0.250
58	54	0.674
59	9	0.761
60	51	0.341
61	39	0.998
62	53	0.893
63	26	0.663

Fig. A.4 Answer to Chap. 2: practice problem #1

Chapter 2: Practice problem #2 answer (see Fig. A.5)

FRAME NO.	Duplicate frame no.	Random number
1	45	0.185
2	102	0.568
3	16	0.700
4	8	0.143
5	109	0.170
6	64	0.403
7	37	0.857
8	31	0.184
9	27	0.459
10	76	0.016
11	9	0.385
12	70	0.741
13	13	0.946
14	32	0.718
15	56	0.784
16	46	0.957
17	3	0.033
18	98	0.677
19	10	0.687
20	100	0.114
21	29	
.		0.501
100	35	0.796
101	20	0.291
102	73	0.408
103	11	0.364
104	24	0.951
105	82	0.482
106	5	0.877
107	17	0.710
108	34	0.578
109	104	0.976
110	51	0.301
111	6	0.171
112	84	0.106
113	96	0.629
114	67	0.008

Fig. A.5 Answer to Chap. 2: practice problem #2

Chapter 2: Practice problem #3 answer (see Fig. A.6)

FRAME NUMBERS	Duplicate frame numbers	Random number
1	47	0.459
2	68	0.252
3	15	0.210
4	69	0.204
5	67	0.116
6	38	0.262
7	43	0.533
8	50	0.361
9	65	0.957
10	40	0.321
11	57	0.351
12	37	0.689
13	22	0.469
14	3	0.889
15	17	0.255
16	60	0.894
17	5	0.545
18	29	0.858
19	74	0.131
20	72	0.322
21	14	0.433
22		·55
	27	0.·
71	46	0.701
72	35	0.460
73	11	0.854
74	7	0.786
75	12	0.547
76	30	0.804

Fig. A.6 Answer to Chap. 2: practice problem #3

Chapter 3: Practice problem #1 answer (see Fig. A.7)

Sample of Kindergarten students for the WPPSI-R IQ test

IQ score				
110	Null hypothesis:	μ	=	110
100				
105				
95	Research hypothesis:	μ	$\neq$	110
98				
102				
105	n	22		
108				
99				
102	Mean	107.64		
104				
98				
112	STDEV	9.45		
105				
98				
104	s.e.	2.02		
115				
118				
120	95% confidence interval			
125				
130	lower limit	103.44		
115				
	upper limit	111.83		

103.44 -------- 107.64 --- 110 ---------- 111.83
lower limit   Mean   Ref.   upper limit
                        value

Result: Since the reference value of 110 is inside the confidence interval, we accept the null hypothesis

Conclusion: This year's kindergarten students have the same IQ Full Scale scores as the average of the kindergarten students over the past three years

Fig. A.7 Answer to Chap. 3: practice problem #1

Chapter 3: Practice problem #2 answer (see Fig. A.8)

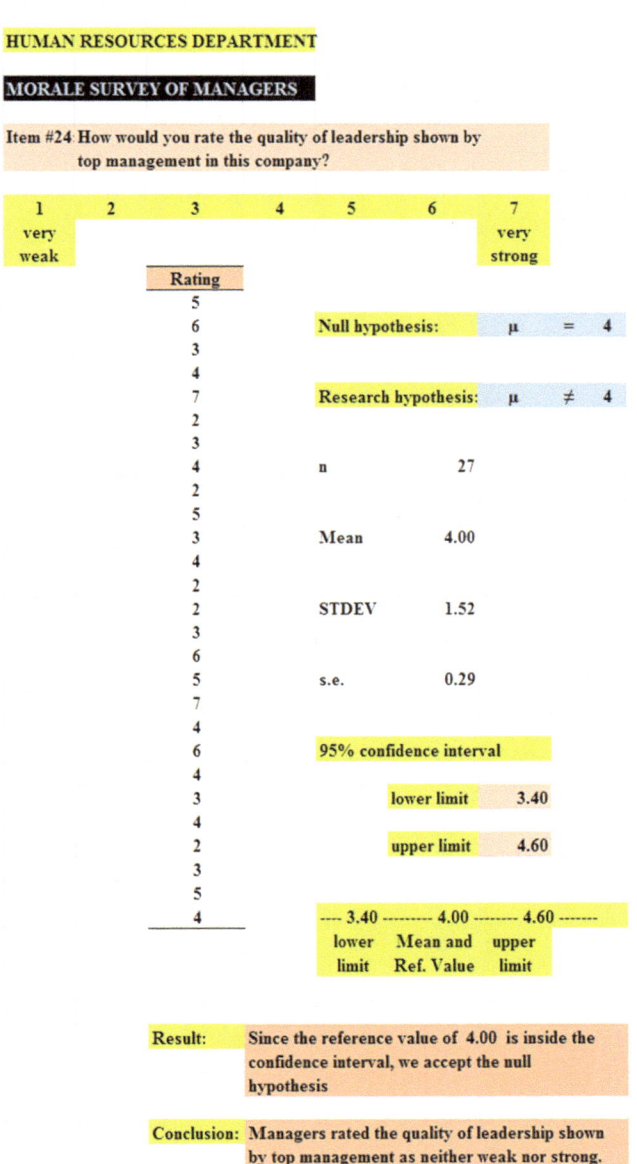

HUMAN RESOURCES DEPARTMENT

MORALE SURVEY OF MANAGERS

Item #24: How would you rate the quality of leadership shown by
top management in this company?

1	2	3	4	5	6	7
very weak						very strong

Rating			
5			
6	Null hypothesis:	μ = 4	
3			
4			
7	Research hypothesis:	μ ≠ 4	
2			
3			
4	n	27	
2			
5			
3	Mean	4.00	
4			
2			
2	STDEV	1.52	
3			
6			
5	s.e.	0.29	
7			
4			
6	95% confidence interval		
4			
3	lower limit	3.40	
4			
2	upper limit	4.60	
3			
5			
4	---- 3.40 --------- 4.00 -------- 4.60 -------		
	lower Mean and upper		
	limit Ref. Value limit		

Result: Since the reference value of 4.00 is inside the
confidence interval, we accept the null
hypothesis

Conclusion: Managers rated the quality of leadership shown
by top management as neither weak nor strong.

Fig. A.8   Answer to Chap. 3: practice problem #2

Chapter 3: Practice problem #3 answer (see Fig. A.9)

FOCUS GROUP PRICING STUDY

Question #10:   "How much would you be willing to pay for this blouse?"

Groups 1, 2, 3 in $		
62	**Null hypothesis:** μ = $68
55	
73	
53	**Research hypothesis:** μ ≠ $68
46	
48	
57	n 30
59	
65	
68	Mean $ 63.23
64	
72	
62	STDEV $ 6.75
67	
59	
71	s.e. $ 1.23
65	
63	
69	**95% confidence interval**
71	
70	lower limit $ 60.71
58	
67	upper limit $ 65.75
65	
63	
59	---- $60.71 --------------- $63.23 ------------- $65.75 ------------- $68 ------
70	lower Mean upper Ref.
67	limit limit Value
64	
65	

Result: Since the reference value is outside of the confidence interval, we reject the null hypothesis and accept the research hypothesis

Conclusion: Adult women (ages 25-44) were willing to pay a price significantly less than $68 , and it was probably closer to $63

Fig. A.9 Answer to Chap. 3: practice problem #3

Chapter 4: Practice problem #1 answer (see Fig. A.10)

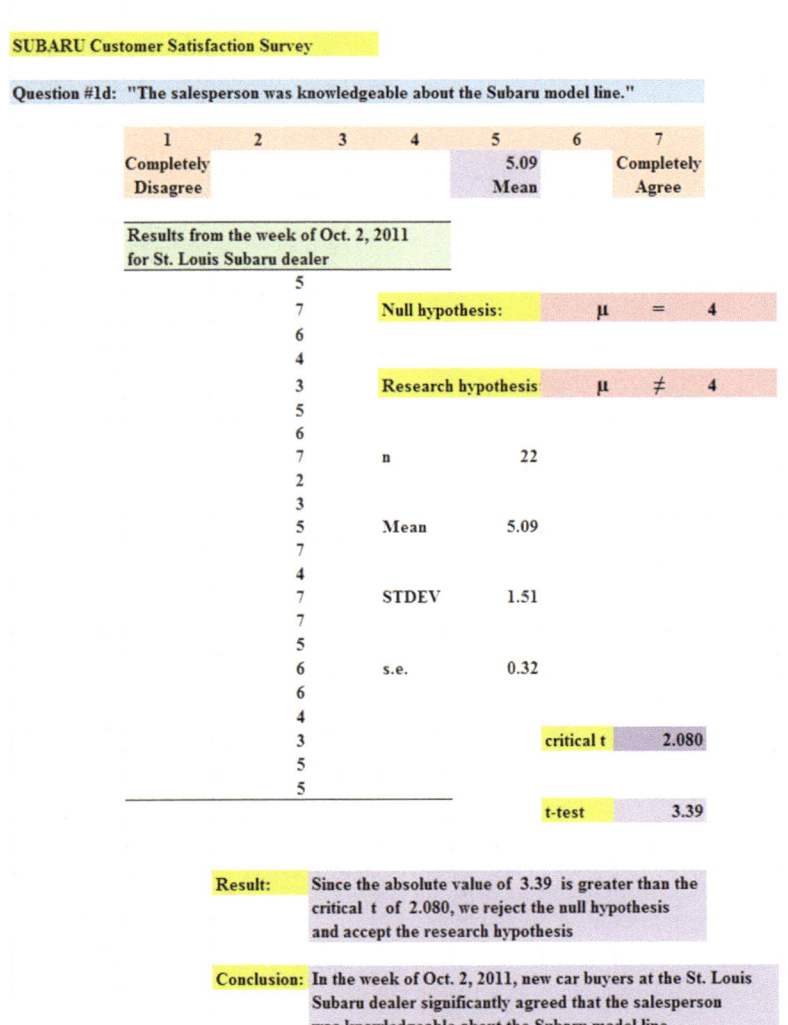

Fig. A.10 Answer to Chap. 4: practice problem #1

Chapter 4: Practice problem #2 answer (see Fig. A.11)

First-semester Calculus
Standardized math test

Score					
87	Null hypothesis:		μ	$=$	88
90					
85					
94	Research hypothesis:		μ	$\neq$	88
93					
88					
82	n	25			
85					
96					
89	Mean	89.96			
92					
94					
91	STDEV	3.55			
89					
90					
92	s.e.	0.71			
93					
89					
88	critical t	2.064			
87					
86					
92	t-test	2.76			
94					
95					
88	Result:	Since the absolute value of 2.76 is greater than the critical t of 2.064, we reject the null hypothesis and accept the research hypothesis			
	Conclusion:	This year's beginning calculus students scored significantly higher on the standardized math test than last year's beginning calculus students.			

Fig. A.11 Answer to Chap. 4: practice problem #2

Chapter 4: Practice problem #3 answer (see Fig. A.12)

MISSOURI BOTANICAL GARDEN

VISITOR SURVEY

Item #10: "How would you rate the helpfulness of The Garden staff?"

1	2	3	4	5	6	7	8	9
poor						6.57		excellent
						Mean		

Results of the week of Nov. 6, 2011

8	
6	Null hypothesis: μ $=$ 5
5	
7	
9	Research hypothesis μ $\neq$ 5
5	
6	
4	n 21
8	
7	
6	Mean 6.57
8	
6	
7	STDEV 1.54
9	
7	
6	s.e. 0.34
3	
8	
7	critical t 2.086
6	
	t-test 4.69

Result: Since the absolute value of 4.69 is greater than the critical value of 2.086, we reject the null hypothesis and accept the research hypothesis

Conclusion: Visitors to the Missouri Botanical Garden during the week of Nov. 6, 2011 rated the helpfulness of The Garden staff as significantly positive

Fig. A.12 Answer to Chap. 4: practice problem #3

Chapter 5: Practice problem #1 answer (see Fig. A.13)

Grade 8 ITBS: Math Problem Solving and Data Interpretation
Grade Equivalent (GE) scores

Group	n	Mean	STDEV
1 Traditional approach	124	8.7	0.8
2 Experimental approach	135	8.9	0.7

Null hypothesis:	$\mu_1 = \mu_2$
Research hypothesis:	$\mu_1 \neq \mu_2$
STDEV1 squared / n1	0.005
STDEV2 squared / n2	0.004
D14 + D17	0.009
s.e.	0.094
critical t	1.960
t-test	-2.133
Result:	Since ther absolute value of - 2.133 is greater than the critical t of 1.96, we reject the null hypothesis and accept the research hypothesis.
Conclusion:	The experimental group had significantly higher grade equivalent scores on the ITBS Math Problem Solving and Data Interpretation test than the traditional group (8.9 vs. 8.7)

Fig. A.13 Answer to Chap. 5: practice problem #1

Chapter 5: Practice problem #2 answer (see Fig. A.14)

Item:	"How interested are you in learning more about how life insurance can provide income for retirement?"

1	2	3	4	5	6	7
Not at all interested		3.44 Women		5.16 Men		Very Interested

Ad: Male model

Men	Women
5	3
6	4
4	6
7	5
5	2
6	3
5	1
4	3
3	2
6	4
7	3
5	5
6	6
4	3
7	4
5	2
4	5
6	3
3	4
7	5
5	4
6	3
2	2
6	4
1	3
7	5
6	1
5	3
4	2
6	3
5	2
7	5
	3
	4

Null hypothesis: $\mu_1 = \mu_2$

Research hypothesis: $\mu_1 \neq \mu_2$

Group	n	mean	STDEV
1 Men	32	5.16	1.51
2 Women	34	3.44	1.31

STDEV1 squared / n1   0.07

STDEV2 squared / n2   0.05

s.e.   0.35

ctitical t   1.96
(df = n1 + n2 -2 = 64)

t-test   4.93

Result: Since the absolute value of 4.93 is greater than the critical t of 1.96, we reject the null hypothesis and accept the research hypothesis

Conclusion: Adult me (ages 25-39) were significantly more interested than adult women (ages 25-39) in learning more about how life insurance can provide income for retirement when a male model was used in the ad (5.16 vs. 3.44)

Fig. A.14 Answer to Chap. 5: practice problem #2

Chapter 5: Practice problem #3 answer (see Fig. A.15)

Grade 3 Iowa Tests of Basic Skills: Vocabulary
Grade Equivalent (GE) scores

Group	n	Mean	STDEV
1 Traditional approach	18	3.7	0.8
2 Experimental approach	19	4.3	0.7

Null hypothesis: $\mu_1 = \mu_2$

Research hypothesis: $\mu_1 \neq \mu_2$

(n1 - 1) x STDEV1 squared	10.88
(n2 - 1) x STDEV2 squared	8.82
n1 + n2 - 2	35
1/n1 + 1/n2	0.11
s.e.	0.25
critical t	2.030
t-test	-2.43

Result:
Since the absolute value of t of - 2.43 is greater than the critical t of 2.030, we reject the null hypothesis and accept the research hypothesis.

Conclusion:
The experimental group had significantly higher grade equivalent (GE) scores in the vocabulary subtest than the traditional group (4.3 vs. 3.7).

Fig. A.15 Answer to Chap. 5: practice problem #3

Chapter 6: Practice problem #1 answer (see Fig. A.16)

SAT REASONING TEST (SAT)

How well can the SAT Critical Reading score predict FROSH GPA?

SAT Critical Reading score	FROSH GPA
250	2.23
400	2.15
600	3.65
500	2.64
520	2.45
720	3.67
630	3.46
550	2.86
650	3.55
630	3.15

correlation	0.88

SUMMARY OUTPUT

Regression Statistics	
Multiple R	0.88
R Square	0.766
Adjusted R Square	0.736
Standard Error	0.305
Observations	10

ANOVA

	df	SS	MS	F	Significance F
Regression	1	2.429	2.429	26.150	0.001
Residual	8	0.743	0.093		
Total	9	3.171			

	Coefficients	Standard Error	t Stat	P-value	Lower 95%	Upper 95%	Lower 95.0%	Upper 95.0%
Intercept	0.920	0.414	2.221	0.057	-0.035	1.876	-0.035	1.876
X Variable 1	0.004	0.001	5.114	0.001	0.002	0.005	0.002	0.005

Fig. A.16 Answer to Chap. 6: practice problem #1

Chapter 6: Practice problem #1 (continued)

1. $a = y\text{-intercept} = 0.920$
2. $b = \text{slope} = 0.004$
3. $Y = a + bX$
   $Y = 0.920 + 0.004\,X$
4. $Y = 0.920 + 0.004\,(600)$
   $Y = 0.920 + 2.4$
   $Y = 3.32$

Chapter 6: Practice problem #2 answer (see Fig. A.17)

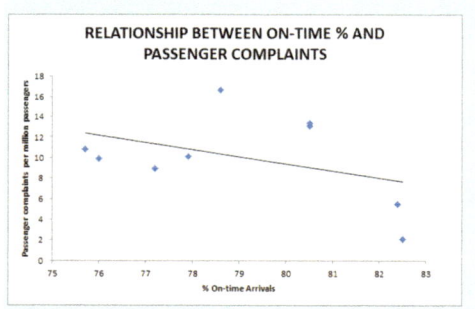

How the Major Airlines Performed in 2009		
Airline	% On-time Arrivals	Passenger complaints per million passengers
Southwest	82.5	2.1
Alaska	82.4	5.5
United	80.5	13.4
US Airways	80.5	13.1
Delta	78.6	16.7
Continental	77.9	10.1
JetBlue	77.2	8.9
AirTran	76.0	9.9
American	75.7	10.8

SUMMARY OUTPUT

Regression Statistics	
Multiple R	0.41
R Square	0.17
Adjusted R Square	0.05
Standard Error	4.23
Observations	9

ANOVA

	df	SS	MS	F
Regression	1	25.05	25.05	1.40
Residual	7	125.51	17.93	
Total	8	150.56		

	Coefficients	Standard Error	t Stat	P-value
Intercept	64.49	46.08	1.40	0.20
X Variable 1	-0.69	0.58	-1.18	0.28

Fig. A.17 Answer to Chap. 6: practice problem #2

Chapter 6: Practice problem #2 (continued)

(2b) About 11 complaints per million passengers

1. $r = -0.41$ (note the negative correlation!)
2. $a = y$-intercept $= 64.49$
3. $b = $ slope $= -0.69$ (note the minus sign as the slope is negative)
4. $Y = a + b X$
   $Y = 64.49 - 0.69 X$
5. $Y = 64.49 - 0.69 (80)$
   $Y = 64.49 - 55.20$
   $Y = 9.29$ complaints per million passengers

Chapter 6: Practice problem #3 answer (see Fig. A.18)

Is there a relationship between absentee rate and Math Test scores?

No. days absent this year	Math Test score
6	92
7	88
21	65
10	75
8	72
9	83
10	86
14	73
16	65
12	71
17	67

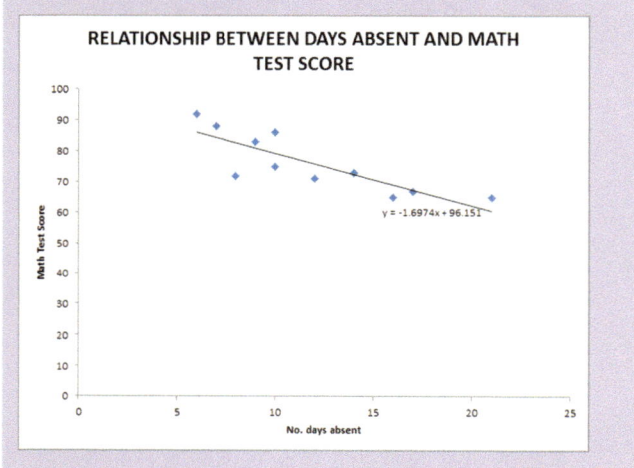

SUMMARY OUTPUT

Regression Statistics	
Multiple R	0.83
R Square	0.686
Adjusted R Square	0.651
Standard Error	5.677
Observations	11

ANOVA

	df	SS	MS	F	Significance F
Regression	1	632.834	632.834	19.635	0.002
Residual	9	290.075	32.231		
Total	10	922.909			

	Coefficients	Standard Error	t Stat	P-value	Lower 95%	Upper 95%	Lower 95.0%	Upper 95.0%
Intercept	96.15	4.840	19.866	0.000	85.203	107.100	85.203	107.100
X Variable 1	-1.70	0.383	-4.431	0.002	-2.564	-0.831	-2.564	-0.831

Fig. A.18 Answer to Chap. 6: practice problem #3

Chapter 6: Practice problem #3 (continued)

1. $r = -0.83$
2. $a = y$-intercept $= 96.15$
3. $b = $ slope $= -1.70$
4. $Y = a + b\,X$
   $Y = 96.15 - 1.70\,X$
5. $Y = 96.15 - 1.70\,(15)$
   $Y = 96.15 - 25.50$
   $Y = 70.65$
   $Y = 71$

Chapter 7: Practice problem #1 answer (see Fig. A.19)

GRADUATE RECORD EXAMINATIONS

How well does the GRE and the GRE subject area test in Psychology predict
GPA at the end of the first year of a Masters' program in Psychology?

FIRST-YEAR GPA	GRE VERBAL	GRE QUANTITATIVE	GRE WRITING	GRE PSYCHOLOGY
3.25	600	620	5	650
3.42	520	550	4	600
2.85	510	540	2	500
2.65	480	460	1	510
3.65	720	710	6	630
3.16	570	610	3	550
3.56	710	650	4	610
2.35	500	480	2	430
2.86	450	470	3	450
2.95	560	530	4	550
3.15	550	580	4	580
3.45	610	620	5	620

SUMMARY OUTPUT

Regression Statistics	
Multiple R	0.93
R Square	0.858
Adjusted R Square	0.777
Standard Error	0.184
Observations	12

ANOVA

	df	SS	MS	F	Significance F
Regression	4	1.436	0.359	10.563	0.004
Residual	7	0.238	0.034		
Total	11	1.674			

	Coefficients	Standard Error	t Stat	P-value	Lower 95%
Intercept	0.5682	0.633	0.898	0.399	-0.928
GRE VERBAL	-0.0004	0.002	-0.211	0.839	-0.005
GRE QUANTITATIVE	0.0022	0.002	0.907	0.394	-0.004
GRE WRITING	0.0501	0.073	0.682	0.517	-0.124
GRE PSYCHOLOGY	0.0024	0.002	1.543	0.167	-0.001

	FIRST-YEAR GPA	GRE VERBAL	GRE QUANTITATIVE	GRE WRITING	GRE PSYCHOLOGY
FIRST-YEAR GPA	1				
GRE VERBAL	0.79	1			
GRE QUANTITATIVE	0.87	0.93	1		
GRE WRITING	0.83	0.74	0.82	1	
GRE PSYCHOLOGY	0.89	0.76	0.83	0.81	1

Fig. A.19 Answer to Chap. 7: practice problem #1

Chapter 7: Practice problem #1 (continued)

1. Multiple correlation $= .93$
2. y-Intercept $= 0.5682$
3. -0.0004
4. 0.0022
5. 0.0501
6. 0.0024
7. $Y = a + b_1 X_1 + b_2 X_2 + b_3 X_3 + b_4 X_4$
   $Y = 0.5682 - 0.0004 X_1 + 0.0022 X_2 + 0.0501 X_3 + 0.0024 X_4$
8. $Y = 0.5682 - 0.0004 (610) + 0.0022 (550) + 0.0501 (3) + 0.0024 (610)$
   $Y = 0.5682 - 0.244 + 1.21 + 0.1503 + 1.464$
   $Y = 3.3925 - 0.244$
   $Y = 3.15$
9. 0.79
10. 0.87
11. 0.83
12. 0.89
13. 0.74
14. 0.76
15. The GRE Psychology Test is the best single predictor of First-Year GPA with a correlation of $+0.89$.
16. The four predictors combined predict the First-Year GPA at $+.93$, and this is slightly better than the best single predictor's correlation of $+.89$.

Chapter 7: Practice problem #2 answer (see Fig. A.20)

NATIONAL FOOTBALL LEAGUE (NFL)		2009 Regular Season			
Team		Offense		Defense	
	Games Won	Yards Gained	Points Scored	Yards allowed	Points allowed
Arizona	10	5510	375	5543	325
Atlanta	9	5447	363	5582	325
Baltimore	9	5619	391	4808	261
Buffalo	6	4382	258	5449	326
Carolina	8	5297	315	5053	308
Chicago	7	4965	327	5404	375
Cincinnati	10	4946	305	4822	291
Cleveland	5	4163	245	6229	375
Dallas	11	6390	361	5054	250
Denver	8	5463	326	5040	324
Detroit	2	4784	262	6274	494
Green Bay	11	6065	461	4551	297
Indianapolis	14	5809	416	5427	307
Jacksonville	7	5385	290	5637	380
Kansas City	4	4851	294	6211	424
Miami	7	5401	360	5589	390
Minnesota	12	6074	470	4888	312
New England	10	6357	427	5123	285
New Orleans	13	6461	510	5724	341
NY Giants	8	5856	402	5198	427
NY Jets	9	5136	348	4037	236
Oakland	5	4258	197	5791	379
Philadelphia	11	5726	429	5137	337
Pittsburgh	9	5941	368	4885	324
San Diego	13	5761	454	5230	320
San Francisco	8	4652	330	5222	281
Seattle	5	5069	280	5703	390
St. Louis	1	4470	175	5965	436
Tempa Bay	3	4600	244	5849	400
Tennesee	8	5623	354	5850	402
Washington	4	4998	266	5115	336
Houston	9	6129	388	5198	333

SUMMARY OUTPUT

Regression Statistics	
Multiple R	0.94
R Square	0.8831
Adjusted R Square	0.8658
Standard Error	1.1807
Observations	32

ANOVA

	df	SS	MS	F	Significance F
Regression	4	284.3581	71.0895	50.9914	3.3639E-12
Residual	27	37.6419	1.3941		
Total	31	322.0000			

	Coefficients	Standard Error	t Stat	P-value	Lower 95%	Upper 95%	ower 95.0%	pper 95.0%
Intercept	0.601	4.0020	0.1503	0.8817	-7.6100	8.8127	-7.6100	8.8127
Yards Gained	0.000	0.0007	0.0633	0.9500	-0.0014	0.0014	-0.0014	0.0014
Points Scored	0.030	0.0056	5.3058	0.0000	0.0181	0.0410	0.0181	0.0410
Yards allowed	0.001	0.0007	1.5543	0.1317	-0.0004	0.0026	-0.0004	0.0026
Points allowed	-0.026	0.0062	-4.2655	0.0002	-0.0389	-0.0136	-0.0389	-0.0136

	Games Won	Yards Gained	Points Scored	Yards allowed	Points allowed
Games Won	1				
Yards Gained	0.77	1			
Points Scored	0.88	0.87	1		
Yards allowed	-0.56	-0.45	-0.47	1	
Points allowed	-0.68	-0.41	-0.46	0.80	1

Fig. A.20 Answer to Chap. 7: practice problem #2

Chapter 7: Practice problem #2 (continued)

1. $R_{xy} = +0.94$
2. y-intercept $= 0.601$
3. Yards gained $= .000$
4. Points scored $= 0.030$
5. Yards allowed $= 0.001$
6. Points allowed $= -0.026$
7. $Y = a + b_1 X_1 + b_2 X_2 + b_3 X_3 + b_4 X_4$
   $Y = 0.601 + 0.000 X_1 + 0.030 X_2 + 0.001 X_3 - 0.026 X_4$
8. $Y = 0.601 + 0.000 (5,100) + 0.030 (360) + 0.001 (5,400) - 0.026 (330)$
   $Y = 0.601 + 0.0 + 10.8 + 5.4 - 8.58$
   $Y = 8.22$
   $Y = 8$ games won
9. $+0.77$
10. $+0.88$
11. -0.56
12. -0.68
13. $+0.87$
14. -0.46
15. The best predictor of games won was points scored with a correlation of $+0.88$.
16. The four predictors combined predict games won with a correlation of $+0.94$ which is much better than the best single predictor by itself.

Chapter 7: Practice problem #3 answer (see Fig. A.21)

Store ID	Y Annual Sales ($000)	X₁ Average Daily Traffic	X₂ Population (2-mile radius)	X₃ Average Income in Area
1	1,121	61,655	17,880	$28,991
2	766	35,236	13,742	$14,731
3	595	35,403	19,741	$8,114
4	899	52,832	23,246	$15,324
5	915	40,809	24,485	$11,438
6	782	40,820	20,410	$11,730
7	833	49,147	28,997	$10,589
8	571	24,953	9,981	$10,706
9	692	40,828	8,982	$23,591
10	1,005	39,195	18,814	$15,703
11	589	34,574	16,941	$9,015
12	671	26,639	13,319	$10,065
13	903	55,083	21,482	$17,365
14	703	37,892	26,524	$7,532
15	556	24,019	14,412	$6,950
16	657	27,791	13,896	$9,855
17	1,209	53,438	22,444	$21,589
18	997	54,835	18,096	$22,659
19	844	32,919	16,458	$12,660
20	883	29,139	16,609	$11,618

SUMMARY OUTPUT

Regression Statistics	
Multiple R	0.91
R Square	0.836
Adjusted R Square	0.806
Standard Error	81.603
Observations	20

ANOVA

	df	SS	MS	F	Significance F
Regression	3	544502.827	181500.942	27.256	1.58756E-06
Residual	16	106544.123	6659.008		
Total	19	651046.950			

	Coefficients	Standard Error	t Stat	P-value	Lower 95%	Upper 95%	Lower 95.0%	Upper 95.0%
Intercept	60.07	91.755	0.655	0.522	-134.440	254.585	-134.440	254.585
Average Daily Traffic	-0.02	0.006	-2.682	0.016	-0.030	-0.004	-0.030	-0.004
Population (2-mile radius)	0.04	0.009	4.401	0.000	0.021	0.060	0.021	0.060
Average Income in Area	0.05	0.010	4.898	0.000	0.028	0.071	0.028	0.071

	Annual Sales ($000)	Average Daily Traffic	Population (2-mile radius)	Average Income in Area
Annual Sales ($000)	1			
Average Daily Traffic	0.77	1		
Population (2-mile radius)	0.42	0.53	1	
Average Income in Area	0.72	0.74	-0.11	1

Fig. A.21 Answer to Chap. 7: practice problem #3

Chapter 7: Practice problem #3 (continued)

1. Multiple correlation $= +.91$
2. y-intercept $= 60.07$
3. Average daily traffic $= -0.02$
4. Population $= 0.04$
5. Average income $= 0.05$
6. $Y = a + b_1 X_1 + b_2 X_2 + b_3 X_3$
   $Y = 60.07 - 0.02 X_1 + 0.04 X_2 + 0.05 X_3$
7. $Y = 60.07 - 0.02 \, (42{,}000) + 0.04 \, (23{,}000) + 0.05 \, (22{,}000)$
   $Y = 60.07 - 840 + 920 + 1{,}100$
   $Y = 1{,}240.07$
   $Y = \$ \, 1{,}240{,}000$ or \$1.24 million
8. $+0.77$
9. $+0.42$
10. $+0.72$
11. $+0.53$
12. -0.11
13. Average daily traffic is the best predictor of annual sales because it has a correlation of $+.77$ with annual sales, and the other two predictors have a correlation that is smaller than 0.77 (0.72 and 0.42).
14. The three predictors combined predict annual sales at $+.91$, and this is much better than the best single predictor's correlation of $+.77$ with annual sales.

Chapter 8: Practice problem #1 answer (see Fig. A.22)

IOWA TESTS OF BASIC SKILLS

GRADE 5 READING COMPREHENSION SCORES

GRADE EQUIVALENT SCORES (GE)

SCHOOL A	SCHOOL B	SCHOOL C
5.2	5.1	5.8
5.5	5.6	5.6
5.9	5.8	6.3
4.6	6.4	5.9
6.2	6.5	6.4
6.4	6.7	6.1
5.8	4.8	6.4
5.9	4.9	6.5
5.4	4.3	6.8
6.3	6.5	7.2
6.4	6.2	6.9
7.3		7.3
		7.4

Anova: Single Factor

SUMMARY

Groups	Count	Sum	Average	Variance
SCHOOL A	12	70.90	5.91	0.48
SCHOOL B	11	62.80	5.71	0.68
SCHOOL C	13	84.60	6.51	0.34

ANOVA

Source of Variation	SS	df	MS	F	P-value	F crit
Between Groups	4.24	2	2.12	4.32	0.02	3.28
Within Groups	16.19	33	0.49			
Total	20.42	35				

SCHOOL A vs. SCHOOL C

1/n SCHOOL A + 1/n SCHOOL C	0.16
s.e. of SCHOOL A vs. SCHOOL C	0.28
ANOVA t-test	-2.14

Fig. A.22 Answer to Chap. 8: practice problem #1

Chapter 8: Practice problem #1 (continued)

1. H_0: $\mu_A = \mu_B = \mu_C$
   H_1: $\mu_A \neq \mu_B \neq \mu_C$
2. 2.12
3. 0.49
4. $F = 4.33$
5. 3.28
6. Result: Since 4.33 is greater than 3.28, we reject the null hypothesis and accept the research hypothesis.
7. There was a significant difference in fifth grade GE reading comprehension scores between the three schools.
   SCHOOL A vs. SCHOOL C
8. H_0: $\mu_A = \mu_C$
   H_1: $\mu_A \neq \mu_C$
9. 5.91
10. 6.51
11. $36 - 3 = 33$
12. Critical $t = 2.035$
13. s.e. $= 0.28$
14. ANOVA $t = -2.14$
15. Result: Since the absolute value of -2.14 is greater than 2.035, we reject the null hypothesis and accept the research hypothesis.
16. Conclusion: The reading teaching method in SCHOOL C had significantly higher GE scores in reading comprehension than the teaching method used in SCHOOL A (6.51 vs. 5.91).
    SCHOOL A vs. SCHOOL B
17. H_0: $\mu_A = \mu_B$
    H_1: $\mu_A \neq \mu_B$
18. 5.91
19. 5.71
20. $36 - 3 = 33$
21. 2.035
22. s.e. $= 0.29$
23. ANOVA $t = 0.68$
24. Result: Since the absolute value of 0.68 is less than 2.035, we accept the null hypothesis.
25. Conclusion: There was no difference in reading comprehension GE scores between SCHOOL A and SCHOOL B.
    SCHOOL B vs. SCHOOL C
26. H_0: $\mu_B = \mu_C$
    H_1: $\mu_B \neq \mu_C$
27. 5.71
28. 6.51
29. $36 - 3 = 33$

Chapter 8: Practice problem #1 (continued)

30. Critical $t = 2.035$
31. s.e $= 0.29$
32. ANOVA $t = -2.79$
33. Result: Since the absolute value of -2.79 is greater than 2.035, we reject the null hypothesis and accept the research hypothesis.
34. Conclusion: The reading teaching method in SCHOOL C had significantly higher GE scores in reading comprehension than the teaching method used in SCHOOL B (6.51 vs. 5.71).
35. Summary: SCHOOL C had significantly higher reading comprehension scores than both SCHOOL A and SCHOOL B. There was no difference in reading scores between SCHOOL A and SCHOOL B.
36. Based on these results, the teaching method used in SCHOOL C resulted in significantly higher reading comprehension scores than the teaching method used in either SCHOOL A or SCHOOL B.

Chapter 8: Practice problem #2 answer (see Fig. A.23)

ANGUS BURGER TEST MARKET STUDY

1	2	3	4
Radio	Local TV	Billboards	Local newspaper
300	310	340	280
320	315	330	285
310	320	345	290
290	326	342	275
280	324	341	282
315	318	351	284
326	330	339	291
295	327	337	284
278	328	329	279
289	319	328	274
287	326	332	283
305	328	335	285

Anova: Single Factor

SUMMARY

Groups	Count	Sum	Average	Variance
Radio	12	3595	299.58	247.54
Local TV	12	3871	322.58	37.72
Billboards	12	4049	337.42	48.63
Local newspaper	12	3392	282.67	26.61

ANOVA

Source of Variation	SS	df	MS	F	P-value	F crit
Between Groups	21172.40	3	7057.47	78.31	1.12354E-17	2.82
Within Groups	3965.42	44	90.12			
Total	25137.81	47				

Billboard ads vs. Radio ads

1/n Billboards + 1/n Radio	0.17
s.e of Billboard ads vs. Radio ads	3.88
ANOVA t-test	9.76

Fig. A.23 Answer to Chap. 8: practice problem #2

Chapter 8: Practice problem #2 (continued)

1. Null hypothesis: $\mu_1 = \mu_2 = \mu_3 = \mu_4$
   Research hypothesis: $\mu_1 \neq \mu_2 \neq \mu_3 \neq \mu_4$
2. $MS_b = 7057.47$
3. $MS_w = 90.12$
4. $F = 78.31$
5. critical $F = 2.82$
6. Since the F-value of 78.31 is greater than the critical F-value of 2.82, we reject the null hypothesis and accept the research hypothesis.
7. There was a significant difference in the number of Angus Burgers sold in the four types of advertising media.
8. Null hypothesis: $\mu_3 = \mu_1$
   Research hypothesis: $\mu_3 \neq \mu_1$
9. 337.42
10. 299.58
11. Degrees of freedom $= 48 - 4 = 44$
12. Critical $t = 1.96$
13. s.e. $_{ANOVA} = SQRT(MS_w \times \{1/12 + 1/12\}) = SQRT (90.12 \times \{0.083+0.083\}$ $= SQRT(14.96) = 3.87$
14. ANOVA $t = (337.42 - 299.58)/3.87 = 9.78$
15. Since the absolute value of 9.78 is greater than the critical t of 1.96, we reject the null hypothesis and accept the research hypothesis.
16. Billboard ads sold significantly more Angus Burgers than Radio ads (337 vs. 300).

Chapter 8: Practice problem #3 answer (see Fig. A.24)

ITEM #8:	"How believable is this commercial to you?"

1	2	3	4	5	6	7	8	9
not very believable								very believable

Rating for Focus Groups 1, 2, 3 combined

	Television commercial		
A	B	C	D
2	3	5	6
3	4	6	7
5	5	7	4
4	2	5	5
5	6	8	3
3	1	6	8
6	4	7	2
4	3	5	6
3	7	4	7
7	6	6	5
2	5	3	8
1	3	6	9
3	4	8	5
5	2	9	6
6	3	5	7

Anova: Single Factor

SUMMARY

Groups	Count	Sum	Average	Variance
A	15	59	3.93	2.92
B	15	58	3.87	2.84
C	15	90	6.00	2.57
D	15	88	5.87	3.70

ANOVA

Source of Variation	SS	df	MS	F	P-value	F crit
Between Groups	62.18	3	20.73	6.89	0.0005	2.77
Within Groups	168.40	56	3.01			
Total	230.58	59				

Commercial B vs. Commercial D	

1/ 15 + 1/ 15	0.13

s.e. ANOVA	0.63

ANOVA t - test	-3.16

Fig. A.24 Answer to Chap. 8: practice problem #3

Chapter 8: Practice problem #3 (continued)

1. Null hypothesis: $\quad\quad\quad\quad \mu_A = \mu_B = \mu_C = \mu_D$
   Research hypothesis: $\quad\quad \mu_A \neq \mu_B \neq \mu_C \neq \mu_D$
2. $MS_b = 20.73$
3. $MS_w = 3.01$
4. $F = 6.89$
5. Critical $F = 2.77$
6. Since the F-value of 6.89 is greater than the critical F-value of 2.77, we reject the null hypothesis and accept the research hypothesis.
7. There was a significant difference in the believability of the four television commercials.
8. Null hypothesis: $\quad\quad\quad\quad \mu_B = \mu_D$
   Research hypothesis: $\quad\quad \mu_B \neq \mu_D$
9. 3.87
10. 5.87
11. Degrees of freedom $= 60 - 4 = 56$
12. Critical $t = 1.96$
13. s.e. $_{ANOVA} = $ SQRT($MS_w \times \{1/15 + 1/15\}$) = SQRT $(3.01 \times \{0.067 + 0.067\})$ = SQRT $(0.40) = 0.64$
14. ANOVA $t = (3.87 - 5.87)/0.64 = -3.125$
15. Since the absolute value of -3.125 is greater than the critical t of 1.96, we reject the null hypothesis and accept the research hypothesis.
16. Commercial D was significantly more believable than Commercial B (5.87 vs. 3.87).

Appendix B: Practice Test

Chapter 1: Practice test

Suppose that you have been asked by the manager of the Webster Groves Subaru dealer in St. Louis to analyze the data from a recent survey of its customers. Subaru of America mails a "SERVICE EXPERIENCE SURVEY" to customers who have recently used the Service Department for their car. Let us try your Excel skills on item #10e of this survey (see Fig. B.1).

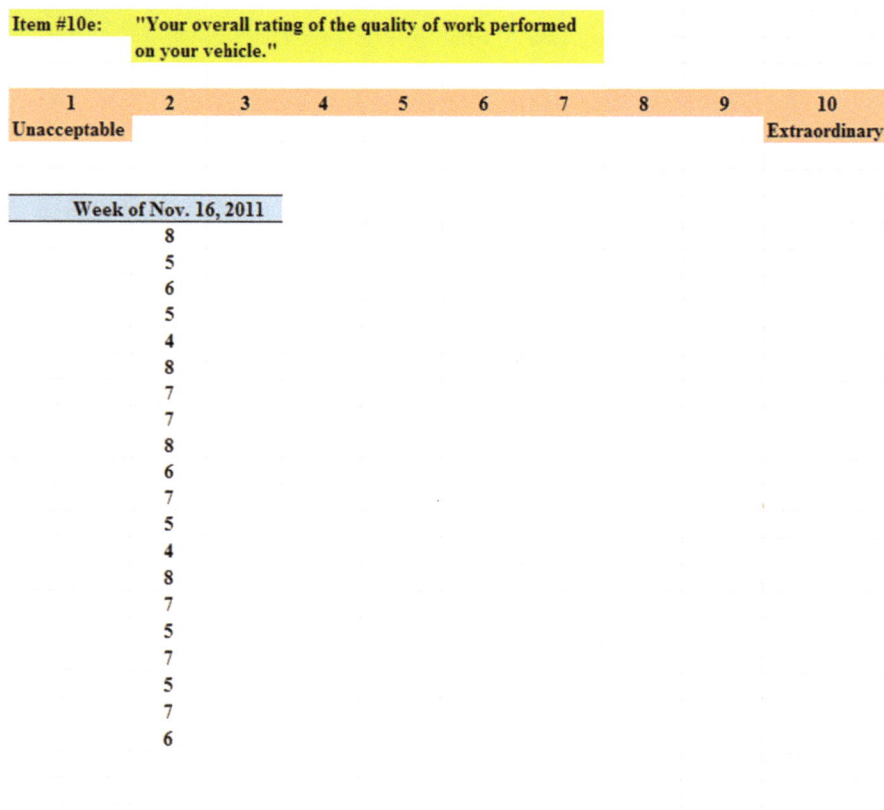

Fig. B.1 Worksheet data for Chap. 1: practice test (practical example)

(a) Create an Excel table for these data and then use Excel to the right of the table to find the sample size, mean, standard deviation, and standard error of the mean for these data. Label your answers and round off the mean, standard deviation, and standard error of the mean to two decimal places.

(b) Save the file as: SUBARU8.

Chapter 2: Practice test

Suppose that you wanted to do a personal interview with a random sample of 12 of a school district's 42 high school science teachers as part of a curriculum revision project.

(a) Set up a spreadsheet of frame numbers for these science teachers with the heading: FRAME NUMBERS

(b) Then, create a separate column to the right of these frame numbers which duplicates these frame numbers with the title: Duplicate frame numbers.

(c) Then, create a separate column to the right of these duplicate frame numbers called RAND NO. and use the $=RAND()$ function to assign random numbers to all of the frame numbers in the duplicate frame numbers column, and change this column format so that 3 decimal places appear for each random number.

(d) Sort the *duplicate frame numbers and random numbers* into a random order.

(e) Print the result so that the spreadsheet fits onto one page.

(f) Circle on your printout the ID number of the first 12 science teachers that you would interview in science curriculum revision project.

(g) Save the file as: RAND15.

Important note. *Note that everyone who does this problem will generate a different random order of teacher ID numbers since Excel assign a different random number each time the RAND() command is used. For this reason, the answer to this problem given in this Excel Guide will have a completely different sequence of random numbers from the random sequence that you generate. This is normal and what is to be expected.*

Chapter 3: Practice test

Webster University, with headquarters in St. Louis, Missouri USA, has over 100 sites where students can take courses, including sites in four European campuses and four sites in Asia for its 21,000 students. Each term, students complete a Course Feedback form at the end of the course, and confidential results are given to the instructors several weeks after the course is completed. Suppose that you have been asked to analyze the data for classes in St. Louis for the previous term, and to test your Excel skills, you have selected a random sample of students from one of the courses. The hypothetical data for item #7 appear in Fig. B.2.

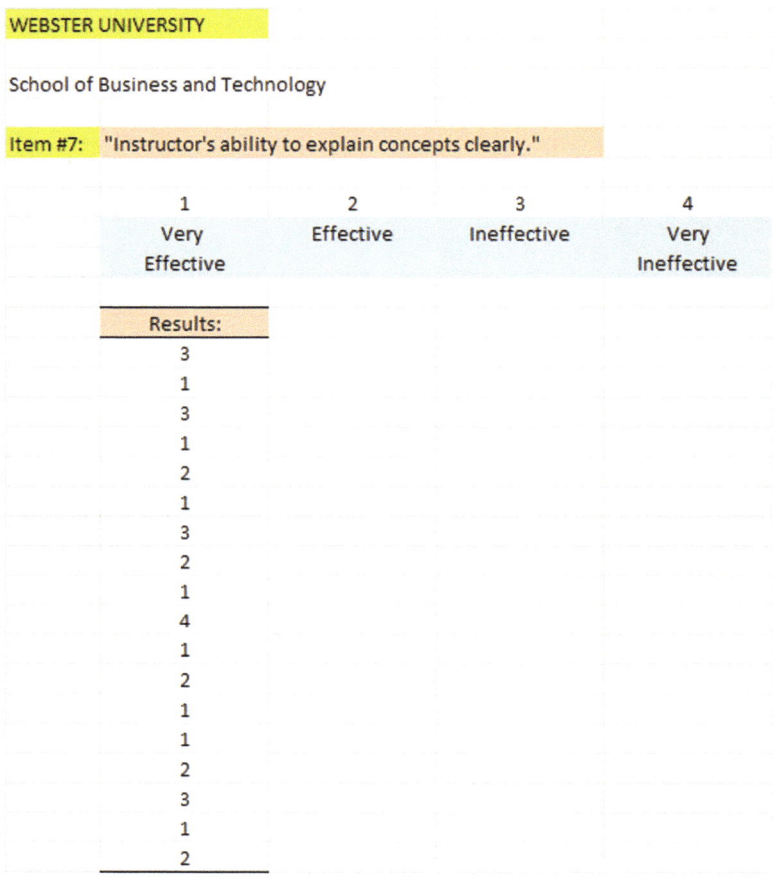

WEBSTER UNIVERSITY

School of Business and Technology

Item #7:   "Instructor's ability to explain concepts clearly."

1	2	3	4
Very Effective	Effective	Ineffective	Very Ineffective

Results:
3
1
3
1
2
1
3
2
1
4
1
2
1
1
2
3
1
2

Fig. B.2 Worksheet data for Chap. 3: practice test (practical example)

(a) Create an Excel table for these data and use Excel to the right of the table to find the sample size, mean, standard deviation, and standard error of the mean for these data. Label your answers and round off the mean, standard deviation, and standard error of the mean to two decimal places in number format.

(b) By hand, write the null hypothesis and the research hypothesis on your printout.

(c) Use Excel's *TINV function* to find the 95% confidence interval about the mean for these data. Label your answers. Use two decimal places for the confidence interval figures in number format.

(d) On your printout, draw a diagram of this 95% confidence interval by hand, including the reference value.

(e) On your spreadsheet, enter the *result*.

(f) On your spreadsheet, enter the *conclusion in plain English*.

(g) Print the data and the results so that your spreadsheet fits onto one page.

(h) Save the file as: Webster5.

Chapter 4: Practice test

Suppose that you have been asked by the American Marketing Association to analyze the data from the 2010 Summer Educators' conference in Boston. In order to check your Excel formulas, you have decided to analyze the data for one of these questions before you analyze the data for the entire survey, one item at a time. The conference used five-point scales with 1 = Definitely Would Not and 5 = Definitely Would. A random sample of the hypothetical data for this one item is given in Fig. B.3.

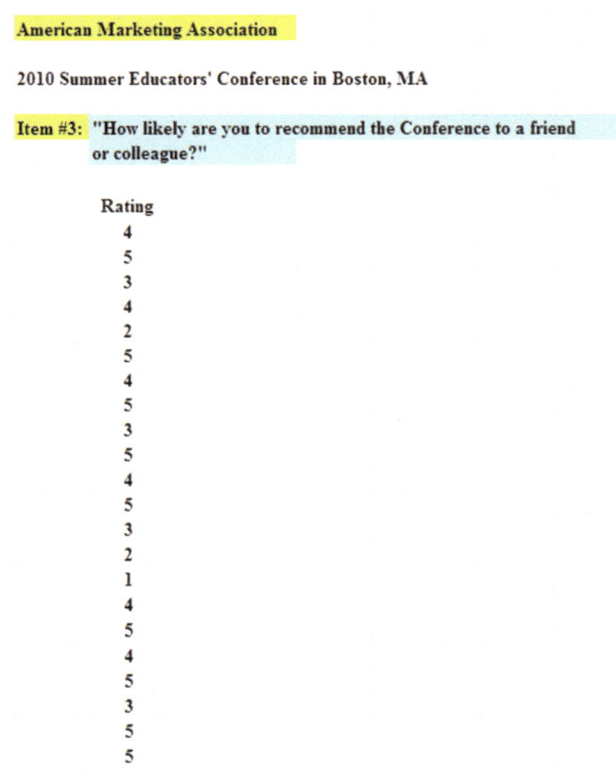

Fig. B.3 Worksheet data for Chap. 4: practice test (practical example)

(a) Write the null hypothesis and the research hypothesis on your spreadsheet.
(b) Create a spreadsheet for these data and then use Excel to find the sample size, mean, standard deviation, and standard error of the mean to the right of the data set. Use number format (three decimal places) for the mean, standard deviation, and standard error of the mean.
(c) Type the *critical t* from the *t*-table in Appendix E onto your spreadsheet and label it.
(d) Use Excel to compute the *t*-test value for these data (use three decimal places) and label it on your spreadsheet.
(e) Type the *result* on your spreadsheet and then type the *conclusion in plain English* on your spreadsheet.
(f) Save the file as: BOS2.

Chapter 5: Practice test

Massachusetts Mutual Financial Group (2010) placed a full page–color ad in *The Wall Street Journal* in which it used a male model hugging a 2-year-old daughter. The ad had the headline and subheadline:

WHAT IS THE SIGN OF A GOOD DECISION?

It's knowing your life insurance can help provide income for retirement. And peace of mind until you get there.

Since the majority of the subscribers to *The Wall Street Journal* are men, an interesting research question would be the following:

Research question: "Does the gender of the model affect adult men's willingness to learn more about how life insurance can provide income for retirement?"

Suppose that you have shown two groups of adult males (ages 25–44), a mock-up of an ad such one group of males saw the ad with a male model, while another group of males saw the identical ad except that it had a female model in the ad. (You randomly assigned these males to one of the two experimental groups.) The two groups were kept separate during the experiment and could not interact with one another.

At the end of a 1-hour discussion of the mock-up ad, the respondents were asked the question given in Fig. B.4.

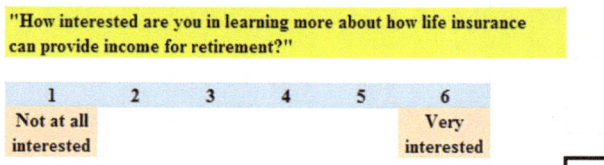

Fig. B.4 Survey item for a mock-up ad (practical example)

The resulting data for this one item appear in Fig. B.5.

MASS MUTUAL FINANCIAL GROUP

Item:	"How interested are you in learning more about how life insurance can provide income for retirement?"				
1	2	3	4	5	6
Not at all interested					Very interested

Male model	Female model
3	4
2	6
4	5
5	3
1	4
6	6
2	6
4	5
3	3
5	5
2	4
4	3
3	5
5	4
1	6
2	5
3	5
1	6
4	4
5	6
6	3
2	4
3	6
1	5
4	6
3	4
5	4

Fig. B.5 Worksheet data for Chap. 5: practice test (practical example)

222   Appendices

(a) Write the null hypothesis and the research hypothesis.
(b) Create an Excel table that summarizes these data.
(c) Use Excel to find the standard error of the difference of the means.
(d) Use Excel to perform a *two-group t-test*. What is the value of *t* that you obtain (use two decimal places)?
(e) On your spreadsheet, type the *critical value of t* using the *t*-table in Appendix E.
(f) Type the *result* of the test on your spreadsheet.
(g) Type your *conclusion in plain English* on your spreadsheet.
(h) Save the file as: lifeinsur3.
(i) Print the final spreadsheet so that it fits onto one page.

Chapter 6: Practice test

Is there a relationship between the amount of money spent on TV ads for a local supermarket chain in St. Louis, Missouri USA and the weekly sales dollars for this chain? You have decided to use the cost of the TV ads as the independent variable (predictor) and weekly sales for the supermarket chain as the dependent variable (criterion). Use simple linear regression for the hypothetical data given in Fig. B.6.

TV ads ($)	Weekly Sales ($)
1,600	24,000
2,400	26,000
2,000	20,000
2,600	23,000
3,400	35,000
2,100	33,000
2,600	24,000
2,700	22,000
2,000	10,000
2,200	31,000
2,600	25,000

Fig. B.6 Worksheet data for Chap. 6: practice test (practical example)

Create an Excel spreadsheet and enter the data:

(a) Create an *XY scatterplot* of these two sets of data such that

- Top title: RELATIONSHIP BETWEEN TV AD COST AND WEEKLY SALES.
- *x*-axis title: TV ads ($).
- *y*-axis title: Weekly sales ($).
- Move the chart below the table.
- Resize the chart so that it is seven columns wide and 25 rows long.
- Delete the legend.
- Delete the gridlines.

(b) Create the *least-squares regression line* for these data on the scatterplot.
(c) Use Excel to run the regression statistics to find the *equation for the least-squares regression line* for these data and display the results below the chart on your spreadsheet. Use number format (two decimal places) for the correlation and for the coefficients.

   Print *just the input data and the chart* so that this information fits onto one page in portrait format.

   Then, print *just the regression output table* on a separate page so that it fits onto that separate page in portrait format.

   By hand:
(d) Circle and label the value of the *y-intercept* and the *slope* of the regression line on your printout.
(e) Write the regression equation *by hand* on your printout for these data (use two decimal places for the *y*-intercept and the slope).
(f) Circle and label the *correlation* between the two sets of scores in the regression analysis summary output table on your printout.
(g) Underneath the regression equation you wrote by hand on your printout, use the regression equation to predict the weekly sales you would expect for a TV ad cost of $2,500 for that week.
(h) *Read from the graph*, the weekly sales you would predict for a TV ad cost of $3,000 and write your answer in the space immediately below.

_____

(i) Save the file as: TV33.

Chapter 7: Practice test

Suppose that you have been hired by an Admissions Director at a local college to analyze the data from the past 3 years to determine if you can predict freshman grade-point average (FROSH GPA) at the end of freshman year by using three predictors for these freshmen: (1) their high school GPA, (2) their SAT-Verbal scores from senior year in high school, and (3) their SAT-Math scores from senior year in high school. The Director wants to use a "rolling average" of the past 3 years of freshmen in the data set. To check your skill in Excel, you have selected a random sample of freshmen from the past 3 years and recorded the information on each student given in the hypothetical table in Fig. B.7.

y	X_1	X_2	X_3
FROSH GPA	HIGH SCHOOL GPA	SAT VERBAL	SAT MATH
3.25	3.35	500	420
2.85	2.93	480	410
3.65	3.25	525	480
3.45	3.35	510	470
3.25	2.85	460	430
2.95	2.75	420	410
2.83	2.58	440	450
2.56	2.66	410	420
3.15	3.25	480	490
3.36	3.42	470	460

Fig. B.7 Worksheet data for Chap. 7: practice test (practical example)

(a) Create an Excel spreadsheet using FROSH GPA as the criterion (Y) and the other variables as the three predictors of this criterion (X_1 = HIGH SCHOOL GPA, X_2 = SAT VERBAL, and X_3 = SAT MATH).
(b) Use Excel's *multiple regression* function to find the relationship between these four variables and place the SUMMARY OUTPUT below the table.
(c) Use number format (two decimal places) for the multiple correlation on the Summary Output, and use three decimal places for the coefficients in the SUMMARY OUTPUT.
(d) Save the file as: GPA5.
(e) Print the table and regression results below the table so that they fit onto one page.

Answer the following questions using your Excel printout:

1. What is the multiple correlation R_{xy}?
2. What is the y-intercept a?
3. What is the coefficient for HIGH SCHOOL GPA b_1?
4. What is the coefficient for SAT VERBAL b_2?
5. What is the coefficient for SAT MATH b_3?
6. What is the multiple regression equation?
7. Predict the FROSH GPA you would expect for high school senior who had a HIGH SCHOOL GPA of 3.15, an SAT VERBAL SCORE of 490, and an SAT MATH score of 480.

(f) Now, go back to your Excel file and create a correlation matrix for these three variables and place it underneath the SUMMARY OUTPUT.
(g) Save this file as GPA6.
(h) Now, print out *just this correlation matrix* on a separate sheet of paper.
Answer the following questions using your Excel printout. (Be sure to include the plus or minus sign for each correlation.):

8. What is the correlation between HIGH SCHOOL GPA and FROSH GPA?
9. What is the correlation between SAT VERBAL and FROSH GPA?

10. What is the correlation between SAT MATH and FROSH GPA?
11. What is the correlation between SAT VERBAL and HIGH SCHOOL GPA?
12. What is the correlation between SAT MATH and HIGH SCHOOL GPA?
13. What is the correlation between SAT MATH and SAT VERBAL?
14. Discuss which of the three predictors is the best predictor of FROSH GPA.
15. Explain in words how much better the three predictor variables combined predict FROSH GPA than the best single predictor by itself.

Chapter 8: Practice test

Suppose that you have been asked to analyze data from a test market study in which three cities with comparable household income levels, population, and other key demographic variables were compared in terms of TV ads run on local channels that stress just one of the following characteristics of the new product in each city: (1) price, (2) quality, and (3) convenience of use. Your task is to determine if there was a significant difference in the number of units sold of this product during test marketing using the hypothetical data in Fig. B.8.

TELEVISION AD EMPHASIS

Price	Quality	Convenience of use
530	180	350
650	210	230
420	275	380
460	275	243
480	340	355
513	250	312
405	250	375
425	225	225
430	224	226
420	275	252
430	255	425
450	220	392
445	260	325
480		334
420		253
410		

Fig. B.8 Worksheet data for Chap. 8: practice test (practical example)

(a) Enter these data on an Excel spreadsheet.
(b) On your spreadsheet, write the null hypothesis and the research hypothesis for these data.
(c) Perform a *one-way ANOVA test* on these data and show the resulting ANOVA table underneath the input data for the three types of TV ad emphases.

(d) If the F-value in the ANOVA table is significant, create an Excel formula to compute the ANOVA *t*-test comparing the number of units sold when price was emphasized vs. the number of units sold when convenience of use was emphasized and show the results below the ANOVA table on the spreadsheet (put the standard error and the ANOVA *t*-test value on separate lines of your spreadsheet and use two decimal places for each value).

(e) Print out the resulting spreadsheet so that all of the information fits onto one page.

(f) On your printout, label by hand the MS (between groups) and the MS (within groups).

(g) Circle and label the value for F on your printout for the ANOVA of the input data.

(h) Label by hand on the printout the mean for price and the mean for convenience of use that were produced by your ANOVA formulas.

(i) Save the spreadsheet as TVad22.
   On a separate sheet of paper, now do the following by hand:

(j) Find the critical value of F in the ANOVA Single Factor results table.

(k) Write a summary of the *result* of the ANOVA test for the input data.

(l) Write a summary of the *conclusion* of the ANOVA test in plain English for the input data.

(m) Write the null hypothesis and the research hypothesis comparing price vs. convenience of use.

(n) Compute the degrees of freedom for the *ANOVA t-test* by hand for three types of ads.

(o) Write the *critical value of t* for the ANOVA *t*-test using the table in Appendix E.

(p) Write a summary of the *result* of the ANOVA *t*-test.

(q) Write a summary of the *conclusion* of the ANOVA *t*-test in plain English.

Reference

Mass Mutual Financial Group. What is the Sign of a Good Decision? (Advertisement) *The Wall Street Journal*, September 29, 2010, p. A22.

Appendix C: Answers to Practice Test

Chapter 1: Practice test answer (see. Fig. C.1)

APPENDIX C

ANSWERS TO PRACTICE TEST

Question #10e: "Your overall rating of the quality of work performed on your vehicle."

Week of Nov. 16, 2011		
8		
5	n	20
6		
5		
4	Mean	6.25
8		
7		
7	STDEV	1.33
8		
6		
7	s.e.	0.30
5		
4		
8		
7		
5		
7		
5		
7		
6		

Fig. C.1 Practice test answer to Chap. 1 problem

Chapter 2: Practice test answer (see. Fig. C.2)

FRAME NUMBERS	Duplicate frame numbers	RAND NO.
1	8	0.335
2	22	0.732
3	31	0.802
4	42	0.163
5	4	0.933
6	29	0.053
7	3	0.379
8	21	0.965
9	37	0.187
10	17	0.523
11	34	0.608
12	25	0.650
13	10	0.705
14	41	0.995
15	30	0.136
16	36	0.501
17	13	0.884
18	15	0.857
19	20	0.709
20	14	0.306
21	9	0.969
22	12	0.282
23	38	0.370
24	26	0.610
25	1	0.215
26	5	0.520
27	35	0.007
28	28	0.368
29	24	0.759
30	32	0.427
31	27	0.373
32	19	0.258
33	6	0.265
34	39	0.018
35	2	0.959
36	18	0.932
37	7	0.311
38	11	0.422
39	16	0.802
40	40	0.710
41	33	0.202
42	23	0.021

Fig. C.2 Practice test answer to Chap. 2 problem

Chapter 3: Practice test answer (see. Fig. C.3)

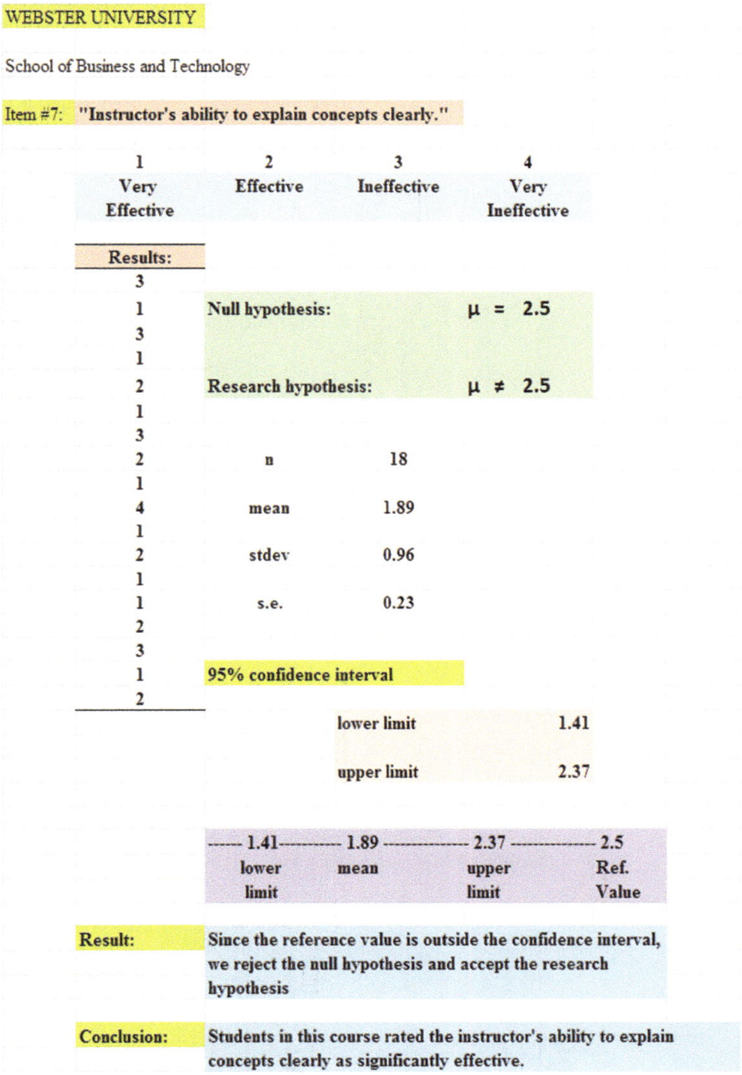

WEBSTER UNIVERSITY

School of Business and Technology

Item #7:   "Instructor's ability to explain concepts clearly."

1	2	3	4
Very Effective	Effective	Ineffective	Very Ineffective

Results:

3		
1	Null hypothesis:	$\mu \ = \ 2.5$
3		
1		
2	Research hypothesis:	$\mu \neq \ 2.5$
1		
3		
2	n	18
1		
4	mean	1.89
1		
2	stdev	0.96
1		
1	s.e.	0.23
2		
3		
1	95% confidence interval	
2		

lower limit	1.41
upper limit	2.37

------ 1.41-----------	1.89 ---------------	2.37 --------------- 2.5
lower limit	mean	upper limit	Ref. Value

Result: Since the reference value is outside the confidence interval, we reject the null hypothesis and accept the research hypothesis

Conclusion: Students in this course rated the instructor's ability to explain concepts clearly as significantly effective.

Fig. C.3   Practice test answer to Chap. 3 problem

Chapter 4: Practice test answer (see. Fig. C.4)

American Marketing Association

2010 Summer Educators' Conference in Boston, MA

Item #3: "How likely are you to recommend the Conference to a friend or colleague?"

Rating		
4	Null hypothesis:	$\mu = 3$
5		
3	Research hypothesis: $\mu \neq 3$	
4		
2	n	22
5		
4		
5	Mean	3.909
3		
5		
4	STDEV	1.192
5		
3		
2	s.e.	0.254
1		
4		
5	critical t	2.080
4		
5		
3	t-test	3.578
5		
5		

Result: Since the absolute value of 3.578 is greater than the critical t of 2.080, we reject the null hypothesis and accept the research hypothesis.

Conclusion: Attendees at the Summer 2010 Educators' Conference of the American Marketing Association in Boston were significantly likely to recommend the Conference to a friend or colleague.

Fig. C.4 Practice test answer to Chap. 4 problem

Chapter 5: Practice test answer (see. Fig. C.5)

MASS MUTUAL FINANCIAL GROUP

Item:	"How interested are you in learning more about how life insurance can provide income for retirement?"

1	2	3	4	5	6
Not at all interested		3.30		4.70	Very interested

Male model	Female model
3	4
2	6
4	5
5	3
1	4
6	6
2	6
4	5
3	3
5	5
2	4
4	3
3	5
5	4
1	6
2	5
3	5
1	6
4	4
5	6
6	3
2	4
3	6
1	5
4	6
3	4
5	4

Group	n	Mean	STDEV
1 Male model	27	3.30	1.54
2 Female model	27	4.70	1.07

Null hypothesis: $\mu_1 = \mu_2$

Research hypothesis: $\mu_1 \neq \mu_2$

$1/n1 + 1/n2$	0.07
$(n1 - 1) \times S1$ squared	61.63
$(n2 - 1) \times S2$ squared	29.63
$n1 + n2 - 2$ (degrees of freedom)	52
s.e.	0.36
critical t	1.96
t-test	-3.90

Result: Since the absolute value of - 3.90 is greater than the critical t of 1.96, we reject the null hypothesis and accept the research hypothesis.

Conclusion: Adult men (ages 25-44) were significantly more interested in learning more about how life insurance can provide income for retirement when a female model was used than when a male model was used in the ad (4.70 vs. 3.30)

Fig. C.5 Practice test answer to Chap. 5 problem

Chapter 6: Practice test answer (see. Fig. C.6)

TV ads ($)	Weekly Sales ($)
1,600	24,000
2,400	26,000
2,000	20,000
2,600	23,000
3,400	35,000
2,100	33,000
2,600	24,000
2,700	22,000
2,000	10,000
2,200	31,000
2,600	25,000

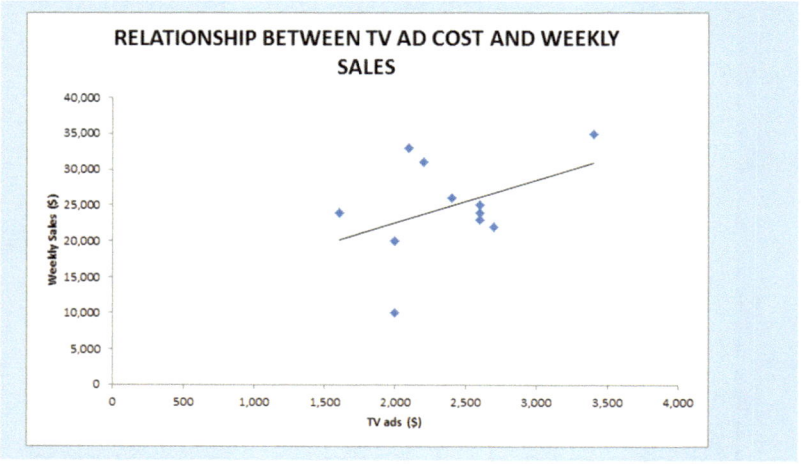

SUMMARY OUTPUT

Regression Statistics	
Multiple R	0.42
R Square	0.175
Adjusted R Square	0.083
Standard Error	6534.831
Observations	11

ANOVA

	df	SS	MS	F	Significance F
Regression	1	81,300,259.12	81,300,259.12	1.903808	0.200962251
Residual	9	384,336,104.51	42,704,011.61		
Total	10	465,636,363.64			

	Coefficients	Standard Error	t Stat	P-value	Lower 95%	Upper 95%	Lower 95.0%	Upper 95.0%
Intercept	10,646.08	10458.50972	1.018	0.335	-13012.71192	34304.87	-13012.71192	34304.87344
X Variable 1	5.95	4.312350048	1.380	0.201	-3.805	15.705	-3.805	15.705

Fig. C.6 Practice test answer to Chap. 6 problem

Chapter 6: Practice test answer (continued)

(d) $a = $ y-intercept $= 10{,}646.08$
  $b = $ slope $= +5.95$
(e) $Y = a + bX$
  $Y = 10{,}646.08 + 5.95\,X$
(f) $r = $ correlation $= +0.42$
(g) $Y = 10{,}646.08 + 5.95\,(2{,}500)$
  $Y = 10{,}646.08 + 14{,}875$
  $Y = \$25{,}521.08$
(h) About \$28,000–\$29,000

Chapter 7: Practice test answer (see. Fig. C.7)

y	X_1	X_2	X_3
FROSH GPA	HIGH SCHOOL GPA	SAT VERBAL	SAT MATH
3.25	3.35	500	420
2.85	2.93	480	410
3.65	3.25	525	480
3.45	3.35	510	470
3.25	2.85	460	430
2.95	2.75	420	410
2.83	2.58	440	450
2.56	2.66	410	420
3.15	3.25	480	490
3.36	3.42	470	460

SUMMARY OUTPUT

Regression Statistics	
Multiple R	0.87
R Square	0.7604
Adjusted R Square	0.6406
Standard Error	0.1979
Observations	10

ANOVA

	df	SS	MS	F	Significance F
Regression	3	0.7457	0.2486	6.3475	0.0272
Residual	6	0.2349	0.0392		
Total	9	0.9806			

	Coefficients	Standard Error	t Stat	P-value	Lower 95%	Upper 95%	Lower 95.0%	Upper 95.0%
Intercept	-0.739	1.0578	-0.6987	0.5109	-3.3273	1.8492	-3.3273	1.8492
HIGH SCHOOL GPA	0.307	0.3582	0.8569	0.4244	-0.5695	1.1833	-0.5695	1.1833
SAT VERBAL	0.004	0.0030	1.4259	0.2038	-0.0031	0.0117	-0.0031	0.0117
SAT MATH	0.002	0.0027	0.7637	0.4740	-0.0045	0.0086	-0.0045	0.0086

	FROSH GPA	HIGH SCHOOL GPA	SAT VERBAL	SAT MATH
FROSH GPA	1			
HIGH SCHOOL GPA	0.79	1		
SAT VERBAL	0.83	0.81	1	
SAT MATH	0.61	0.54	0.54	1

Fig. C.7 Practice test answer to Chap. 7 problem

Chapter 7: Practice test answer (continued)

1. $R_{xy} = +0.87$
2. y-intercept $= -0.739$
3. $b_1 = 0.307$
4. $b_2 = 0.004$
5. $b_3 = 0.002$
6. $Y = -0.739 + 0.307\ X_1 + 0.004\ X_2 + 0.002\ X_3$
7. $Y = a + b_1\ X_1 + b_2\ X_2 + b_3\ X_3$
   $Y = -0.739 + 0.307\ (3.15) + 0.004\ (490) + 0.002\ (480)$
   $Y = -0.739 + 0.967 + 1.96 + 0.96$
   $Y = 3.15$
8. +.79
9. +.83
10. +.61
11. +.81
12. +.54
13. +.54
14. The best predictor of FROSH GPA was SAT V with a correlation of +.83.
15. The three predictors combined predict FROSH GPA with a correlation of +.87 which is slightly better than the best single predictor by itself.

Chapter 8: Practice test answer (see. Fig. C.8)

TELEVISION AD EMPHASIS

Price	Quality	Convenience of use
530	180	350
650	210	230
420	275	380
460	275	243
480	340	355
513	250	312
405	250	375
425	225	225
430	224	226
420	275	252
430	255	425
450	220	392
445	260	325
480		334
420		253
410		

Anova: Single Factor

SUMMARY

Groups	Count	Sum	Average	Variance
Price	16	7368	460.50	3885.33
Quality	13	3239	249.15	1565.97
Convenience of use	15	4677	311.80	4659.89

ANOVA

Source of Variation	SS	df	MS	F	P-value	F crit
Between Groups	349156.09	2	174,578.04	50.30	9.24E-12	3.23
Within Groups	142310.09	41	3,470.98			
Total	491466.18	43				

Price vs. Convenience of Use

1/n1 + 1n2	0.13
s.e.	21.17
ANOVA t-test	7.02

Fig. C.8 Practice test answer to Chap. 8 problem

(b) $H_0 : \mu_P = \mu_Q = \mu_{CU}$
　　$H_1 : \mu_P \neq \mu_Q \neq \mu_{CU}$
(f) $MS_b = 174{,}578.04$ and $MS_w = 3{,}470.98$
(g) $F = 174{,}578.04/3{,}470.98 = 50.30$
(h) Mean of price $= 460.50$ and Mean of Convenience of Use $= 311.80$
(j) Critical $F = 3.23$

Chapter 8: Practice test answer (continued)

(k) Result: Since 50.30 is greater than 3.23, we reject the null hypothesis and accept the research hypothesis.

(l) Conclusion: There was a significant difference in the number of units sold between the three types of TV ads.

(m) H_0: $\mu_P = \mu_{CU}$

H_1: $\mu_P \neq \mu_{CU}$

(n) df $= n_{TOTAL} - k = 44 - 3 = 41$

(o) Critical $t = 1.96$

(p) Result: Since the absolute value of 7.02 is greater than the critical t of 1.96, we reject the null hypothesis and accept the research hypothesis.

(q) TV ads emphasizing price sold significantly more units than TV ads emphasizing convenience of use (460 vs. 312).

Appendix D: Statistical Formulas

Mean $\bar{X} = \frac{\Sigma X}{n}$

Standard deviation $\text{STDEV} = S = \sqrt{\frac{\Sigma(X-\bar{X})^2}{n-1}}$

Standard error of the mean $\text{s.e.} = S_{\bar{X}} = \frac{S}{\sqrt{n}}$

Confidence interval about the mean $\bar{X} \pm t\, S_{\bar{X}}$

                                                        where $S_{\bar{X}} = \frac{S}{\sqrt{n}}$

One-group t-test $t = \frac{\bar{X} - \mu}{S_{\bar{X}}}$

                                                        where $S_{\bar{X}} = \frac{S}{\sqrt{n}}$

Two-group t-test:

(a) When both groups have a sample size greater than 30

$$t = \frac{\bar{X}_1 - \bar{X}_2}{S_{\bar{X}_1 - \bar{X}_2}}$$

where $S_{\bar{X}_1 - \bar{X}_2} = \sqrt{\frac{S_1^2}{n_1} + \frac{S_2^2}{n_2}}$

and where $df = n_1 + n_2 - 2$

(b) When one or both groups have a sample size less than 30

$$t = \frac{\bar{X}_1 - \bar{X}_2}{S_{\bar{X}_1 - \bar{X}_2}}$$

where $S_{\bar{X}_1 - \bar{X}_2} = \sqrt{\frac{(n_1-1)S_1^2 + (n_2-1)S_2^2}{n_1+n_2-2}\left(\frac{1}{n_1} + \frac{1}{n_2}\right)}$

and where $df = n_1 + n_2 - 2$

Correlation $r = \frac{\frac{1}{n-1}\Sigma(X-\bar{X})(Y-\bar{Y})}{S_x S_y}$

                                                        where S_x = standard deviation of X
                                                        and where S_y = standard deviation of Y.

Simple linear regression $Y = a + bX$
                                                        where a = y-intercept and b = slope of the line.

Multiple regression equation $Y = a + b_1 X_1 + b_2 X_2 + b_3 X_3 +$ etc.
                                                        where a = y-intercept.

One-way ANOVA F-test $F = \mathrm{MS_b/MS_w}$

ANOVA t-test $\mathrm{ANOVA}\ t = \frac{\bar{X}_1 - \bar{X}_2}{\mathrm{s.e._{ANOVA}}}$

where $\mathrm{s.e._{ANOVA}} = \sqrt{MS_w \left(\frac{1}{n_1} + \frac{1}{n_2} \right)}$

and where $\mathrm{df} = n_{\mathrm{TOTAL}} - k$

where $n_{\mathrm{TOTAL}} = n_1 + n_2 + n_3 +$ etc.

and where $k =$ the number of groups.

Appendix E: *t*-Table

Critical t-values needed for rejection of the null hypothesis (see Fig. E.1)

sample size n	degrees of freedom df	critical t
10	9	2.262
11	10	2.228
12	11	2.201
13	12	2.179
14	13	2.160
15	14	2.145
16	15	2.131
17	16	2.120
18	17	2.110
19	18	2.101
20	19	2.093
21	20	2.086
22	21	2.080
23	22	2.074
24	23	2.069
25	24	2.064
26	25	2.060
27	26	2.056
28	27	2.052
29	28	2.048
30	29	2.045
31	30	2.042
32	31	2.040
33	32	2.037
34	33	2.035
35	34	2.032
36	35	2.030
37	36	2.028
38	37	2.026
39	38	2.024
40	39	2.023
infinity	infinity	1.960

Fig. E.1 Critical *t*-values needed for rejection of the null hypothesis

Index